T0305707

Natural HDAC Inhibitors for Epigenetic Combating of Cancer Progression

Natural HDAC Inhibitors for Epigenetic Combating of Cancer Progression deals only with HDAC inhibitors from natural origins including bacteria, fungi, marine organisms and, notably, from diverse plant sources. This book is unique in the sense that it is the only book that discusses wholly and solely HDAC inhibitors of natural origin in the context of cancer chemotherapy. Another peculiar feature of this book is that it debates futuristic nanotechnology approaches for escalating the aqueous solubility, cancer cell uptake, bioavailability and other favourable pharmacological parameters, including the cytotoxicity of natural HDAC inhibitors against cancer cells.

The major features of this book encompass

- General compendium of HDAC inhibitors with a deep emphasis on the toxicity issues of synthetic HDAC inhibitors
- Various groups of natural HDAC inhibitors, their representatives and premier sources
- Cyclic tetrapeptides of natural origin and their importance as cancer chemotherapeutic agents
- Hydroxamates and depsipeptides from natural sources and their promising role in cancer therapy
- Natural flavonoids, their HDAC inhibitory tendency and marvellous anticancer activity
- Non-flavonoid natural HDAC inhibitors and their pleasing cytotoxic effects towards cancer models
- Combined therapy involving natural flavonoids with other anticancer molecules for synergistic and additive benefits against cancer models
- Non-flavonoid HDAC inhibitors and conventional drugs in collaborative mode against aggressive malignancies
- Nanotechnology-based delivery of natural HDAC inhibitors for greater therapeutic efficacy over traditional combinatorial therapy

This book is highly beneficial to university professors and research scholars working on epigenetic therapeutics in general, and natural HDAC inhibitors in particular. This book is equally important to medical oncologists, biochemistry as well as pharmacy candidates and students of master's and undergraduate level with a desire to do a doctorate on HDACs, natural HDAC inhibitors, HDAC inhibitor (natural)-based combinatorial chemotherapy and the delivery of these inhibitors selectively to tumour sites through revolutionary nanotechnological tactics.

Natural HDAC Inhibitors for Epigenetic Combating of Cancer Progression

Shabir Ahmad Ganai

CRC Press
Taylor & Francis Group
Boca Raton London New York

CRC Press is an imprint of the
Taylor & Francis Group, an **informa** business

First edition published 2023
by CRC Press
6000 Broken Sound Parkway NW, Suite 300, Boca Raton, FL 33487-2742

and by CRC Press
4 Park Square, Milton Park, Abingdon, Oxon, OX14 4RN

© 2023 Shabir Ahmad Ganai

CRC Press is an imprint of Taylor & Francis Group, LLC

ISBN: 978-1-032-27986-2 (HB)
ISBN: 978-1-032-27988-6 (PB)
ISBN: 978-1-003-29486-3 (EB)

DOI: 10.1201/9781003294863

Typeset in Times
by Newgen Publishing UK

I dedicate this book to my parents, wife, kids,
Mohammad Wildan Ganai, Asma and Shifa

Contents

Foreword

Functional genomics studies and cancer cell line sensitivity profiling are consistently pointing toward transcriptional and epigenetic networks as potential opportunities for therapeutic modulation. Indeed, the oncology community recognized the crucial roles for histone deacetylases, enzymes that remove acetyl groups from N-acetyl lysine amino acids on histone and non-histone substrates, in regulating a variety of critical cancer-related transcriptional and epigenetic processes long before the era of cancer dependency mapping. As such, a diversity of approaches and strategies have been used to identify potent inhibitors of HDAC activity with the goals of developing tools for fundamental mechanistic biology as well as potential therapeutics.

The current volume is a *tour de force* description of HDAC inhibitors of natural origin. The author diligently describes the overall history of HDAC inhibitor development, paying careful attention to the challenges of both synthetic and natural molecules and to the challenges associated with substantial off-target effects associated with synthetic molecules. This description sets the scene for a deep dive into the world of natural product HDAC inhibitors, including molecules from plant and marine sources. The volume covers all the critical natural chemotypes that inspire new efforts to develop next-generation, synthetic molecules. The chapters are logically organized around these hallmark classes of molecules. Critically, the book balances description of the discovery and fundamental biochemical properties of these inhibitors with details about the antineoplastic and pharmacologic activities of the compounds, as single agents or combinations with other drugs that can impinge on common networks. Additionally, the volume concludes by describing more recent efforts to integrate nanotechnology-based approaches to impact the targeted delivery and pharmacokinetics associated with natural product-based HDAC inhibitors. The final chapter also includes a forward-looking perspective on nano-combinatorics as a therapeutic strategy in oncology.

The author has done an incredible job of providing a thorough and up-to-date compendium of this expansive field. Indeed, the volume is appealing to me as a teacher, given that the structure, organization, and detail matches the organization of a course on this subject. As such, I expect that the volume will be an excellent tool for the classroom. By describing both the history of the field and the state-of-the art in the field, the volume is likely to inspire a new generation to innovate in the area of histone deacetylase inhibition.

It is an honour to provide the foreword for this book. I am most pleased to praise the author on this impressive volume, and to encourage the reader to join in on the quest to develop epigenetic therapies for oncology and beyond!

Angela N. Koehler
Kathleen and Curtis Marble Professor, Biological Engineering
Massachusetts Institute of Technology, USA
August 2022

Preface

After writing two timely books on histone deacetylase inhibitors (HDACi), one in the context of neuroscience and the other keeping cancer in focus, the author felt that there is still a big lacuna, as no comprehensive book was available that has solely discussed HDACi from natural sources, such as marine organisms, fungi, bacteria and, most importantly, from plant sources in relation to their activity against cancer. Keeping this entire scenario in view, the author committed and prepared himself to write a book that would discuss natural HDACi right from the beginning. While working on synthetic HDACi, the author encountered certain research findings highlighting the significant clinical toxicities of these inhibitors. It was the turning point when the author diverted his focus towards HDACi from plant sources. Within a brief time frame the author published original research articles and some mindblowing review articles on typical plant-derived inhibitors, namely apigenin, luteolin and sulforaphane –N-acetyl-cysteine. chrysin and sulforaphane. Further, exploration indicated that the free forms of natural HDACi do not yield the desired outcomes and encapsulating these natural agents singly or in association with standard anticancer therapeutics has resulted in augmented pharmacological advantage.

All these novel research findings motivated the author to write a detailed book on natural HDACi. This book explains the classification of natural HDACi and gives a deep overview of various sources of these inhibitors. Following this, the anticancer benefits of natural cyclic tetrapeptide inhibitors are unbossomed along with the downstream molecular mechanisms elicited. Further, the author copiously describes natural hydroxamates and depsipeptides in the context of distinct malignancies. The abundance of plant-isolated HDACi compelled the author to discuss these inhibitors in four tandem chapters. In one chapter, a thorough classification of flavonoid HDACi and their promising anticancer effects as single agents are elaborated. This is followed by discussion on non-flavonoid natural HDACi as individual anticancer agents. Then the combinatorial therapy of flavonoid HDACi with standard cancer chemotherapeutics for gaining additional therapeutic advantage is summarized. Similarly non-flavonoid natural HDACi are intensely discussed in the context of combinatorial anticancer therapy. Notably, in the closing segment of the book nanotechnology-based approaches for enhancing the efficacy of HDACi are illuminated. This chapter discusses plant-based HDACi in nano-encapsulated form and in nano-combinatorial form against multiple cancers.

This book is important for medical doctors, university students, research scholars, scientists working on epigenetics and epidrugs, especially HDACi.

Shabir

Dr Shabir Ahmad Ganai
Assistant Professor
Division of Basic Sciences & Humanities
FoA, Sopore -193201, SKUAST-Kashmir
NET;JK-SET; GATE-2020; GATE-2022

Author Biography

Dr Shabir Ahmad Ganai completed his master's in Biochemistry at Bangalore University. Following this, he did his PhD in Biochemistry in Tamil Nadu. During his doctorate, Dr Ganai remained AYUSH-JRF. Further, Dr Ganai worked as a Research Associate at the National Brain Research Centre, Haryana. Afterwards, Dr Ganai by competing at national level, got a large grant from the Science and Engineering Research Board and served as Principal Investigator in the University of Kashmir. Dr Ganai currently serves the SKUAST-Kashmir. Dr Ganai has cleared the National Eligibility Test (NET) exam, Jammu and Kashmir State Eligibility Test (JK-SET), GATE Life Sciences Two times (GATE XL 2020 and GATE XL-2022). Dr Ganai has received a couple of best oral presentation awards for presenting his research work in various conferences.

Dr Shabir Ahmad Ganai has published 29 international articles in highly reputed SCIE journals, two international books in biomedical sciences for Springer-Nature and one national level book on structural biochemistry (drug designing). The current cumulative impact factor of articles published by Dr Ganai is 151. Dr Ganai currently has 727 citations, an h-index of 17 and an i10-index of 22. Dr Ganai serves as a referee for multiple international journals of high repute. Dr Ganai has received a certificate of appreciation from the American Chemical Society. Dr Ganai is a member of the American Chemical Society and the Epigenetics Society. Most notably, Dr Shabir Ahmad Ganai figured consecutively in the list of the top 2% scientists of the world, the ranking undertaken by Stanford University experts and Elsevier BV.

1 Abridgement of Classical Histone Deacetylases, Their Inhibitors and Jeopardy of Synthetic Histone Deacetylase Inhibitors

Enzymes transforming the chromatin compactness and ensuing gene expression are emerging as promising targets for therapeutic modalities. Family of such enzymes called histone deacetylases (HDACs), induce the transcriptionally silent state of chromatin through deacetylation of DNA-associated histones. Being antagonistic to histone acetyltransferases (HATs) in function, these enzymes perpetuate the acetylation homeostasis critical for accurate cellular functioning. Any abnormality in expression of HDACs ends up in cancer. Agents fitting into the specific pocket of HDACs thereby impeding their activity have sprouted as encouraging anticancer therapeutics. These agents, known as histone deacetylase inhibitors (HDACi), being natural or synthetic, have their own advantages and shortcomings. Here the different HDAC classes, with the main focus on classical HDACs, the various types of inhibitors of classical HDACs and the substantial off-target effects of synthetic HDACi, have been copiously described.

1.1 GLIMPSE OF HATS AND HDACS

It is well accepted that post-translational modifications of proteins beneath DNA have substantial influence on chromatin conformation and upcoming gene expression (Gibney & Nolan 2010). While few modifications, such as histone methylation, modulate gene expression in a degree-dependent and site-specific manner, histone acetylation unanimously recommends transcriptional activation (Greer & Shi 2012; Ganai 2020a). The negative charge on the acetyl-group advocates histone-DNA repulsion, which in turn gives rise to loose chromatin topology and, in the long run, gene expression. Inversely, the removal of this acetylation tag augments histone-DNA attraction yielding compact chromatin structure, ultimately stimulating transcriptional switch off. In human beings the acetylation balance or homeostasis is under the influence of two antagonistically acting enzyme families. These families engirdle histone acetyltransferases (HATs) and histone deacetylases (HDACs). The

DOI: 10.1201/9781003294863-1

premier function of HATs is to acetylate the DNA underlying histones, although they can perform such activity even on non-histone players. The process of acetylation by HATs utilizes acetyl coenzyme A (Acetyl-CoA) as a cofactor. Enzymes that undo histone acetylation are known as HDACs (Yang & Seto 2007). These enzymes act on acetylated histones and reverse the HAT-mediated acetylation (Ganai 2020b). From the above discussion one thing is obvious that HATs create a conducive environment for transcriptional events due to their chromatin decondensation role, while HDACs make the ground uncomplimentary for gene expression because of their chromatin condensation effect. The typical activity of antiparallel HATs and HDACs plays a key role in upkeeping acetylation homeostasis (Saha & Pahan 2006; Sheikh 2014; Richters & Koehler 2017). It has been attested that peculiar HDAC expression elicits distortion of this homeostasis and ends up in dysregulation of transcription. This dysregulation results in abnormal levels of transcripts and protein levels of certain cancer-driving genes (Hai et al. 2021).

1.2 FAMILY OF CLASSICAL HDACs

HDAC enzymes subdue gene expression through multiple pathways. They regulate the expression of many genes by interacting with transcription factors directly. These transcription factors include p53, Stat3, NF-κB and retinoblastoma protein (Lin et al. 2006; Ropero & Esteller 2007). They also regulate gene expression by giving rise to corepressor complexes with nuclear receptor. Another way by which HDACs regulate differentiation, cell cycle progression and apoptosis is through deacetylation of non-histone proteins (Minucci & Pelicci 2006; Ganai 2018). HDACs require nicotine adenine dinucleotide (NAD^+) or zinc as a cofactor. The former HDACs are known as sirtuins, while those backed by zinc are termed classical HDACs. In humans, classical HDACs are 11 in number and come under the banner of three separate classes. Class I occupies four HDACs (HDAC1–3 and HDAC8), while Class II has total 6 HDACs under its confines among which four are Class IIa ((HDAC4, 5, 7, 9) members and two come under the boundary of Class IIb (HDAC6 and HDAC10). A unique HDAC (HDAC11) that is poorly explored comes under the umbrella of Class IV (Park & Kim 2020). Class III HDACs are not regarded as classical HDACs but sirtuins, as they differ from classical HDACs in many aspects, the critical one being the diverse cofactor requirement (Figure 1.1) (Ganai 2020b). While Class I HDACs are spaced principally in cell nucleus, Class II HDACs occur in cytoplasm as well in nucleus (Bannister & Kouzarides 2011; Rajan, Shi & Xue 2018). In a nutshell, classical HDACs are considered as zinc-reliant amidohydrolases. Over expression of one or more isozymes of zinc-driven HDACs has been reported in several cancers, such as prostate, pancreatic, lung, breast and many more (Glozak & Seto 2007; Ganai 2020c).

1.3 CONCISE EXPLANATION OF HDACi AND THEIR DIVERSE GROUPS

The activity of histone deacetylases can be stopped by natural or synthetic molecules termed HDAC inhibitors (HDACi). The major proportion of these inhibitors

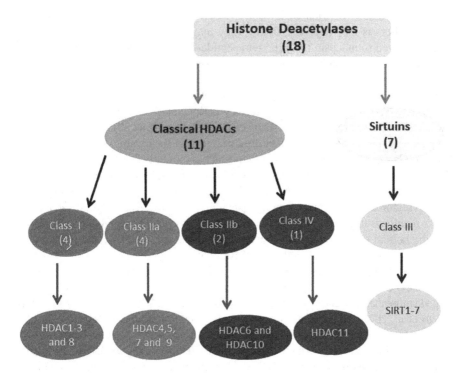

FIGURE 1.1 Arrangement of 18 human HDACs into different classes. HDACs are broadly borderlined into classical HDACs and sirtuins. Classical HDACs form a broader group and are dependent on zinc while sirtuins rely on NAD +. Classical HDACs are further subdivided into three classes ranging from Class I–II and Class IV. Four HDACs come under Class I, four fall within Class IIa, two under Class IIb and one HDAC falls within the roof of Class IV. Sirtuins total seven in number and are named SIR1–SIR7.

encompassing trichostatin A (TSA), vorinostat, dacinostat and panobinostat hamper the activity of HDACs reversibly, while some inhibitors thwart HDACs in a covalent fashion. Among the covalent HDACi, trapoxins and depudecin are notable (Salvador & Luesch 2012). It has been deciphered that histone deacetylase inhibitors restore typical gene expression by reinstating acetylation uniformity (Rossaert et al. 2019). Based on chemotype difference, HDACi have been put into multiple groups. The massive group is hydroxamates, which includes a large number of HDACi. Vorinostat, the primary HDACi that gained approval from the FDA for treatment of cutaneous-T cell lymphoma (CTCL), falls in this group. Belinostat, another member of this group, was granted approval by FDA for peripheral T-cell lymphoma (PTCL). Panobinostat is the most recent HDAC inhibitor to gain FDA approval for use in multiple myeloma subjects. Hydroxamates have stronger inhibitory potential as compared to other groups of HDACi. The second group of HDACi is short chain fatty acids, which covers multiple members including butyrate. Short chain fatty acids have manifested least strength in preclinical trials. Another group is benzamide derivatives, and this group includes tacedinaline and two more members (Li, Tian & Zhu 2020). Apart from these groups, another group of such inhibitors is macrocyclic HDACi. They may be bicyclic depsipeptides such as FDA-approved romidepsin or cyclic tetrapeptides encompassing trapoxin A, apicidin and HC-toxin. Thus, one thing is sure; that among the four FDA-stamped inhibitors, three belong to hydroxamate group and one to bicyclic depsipeptides (Mwakwari et al. 2010). Chidamide was approved by China's national medical and products administration (NMPA) for recurrent and refractory peripheral T-cell lymphoma, which was further extended to breast cancer (Zhang et al. 2018b; Sun et al. 2019). Certain HDACi, such as pracinostat, received orphan drug designation for treating acute myeloid leukaemia (Garcia-Manero et al. 2019). Givinostat has currently been granted approval by the FDA for paediatric Duchenne Muscular Dystrophy (DMD) (Licandro et al. 2021). Three different subunits occurring in tandem exist in a typical HDAC inhibitor. While the cap region evinces contact with the gate amino acid residues of HDAC active site, linker manifests interactions with tube residues of groove. The zinc binding group fits into the bottom of the functional cavity, where it interacts with zinc and other crucial residues (Noureen, Rashid & Kalsoom 2010; Zhang et al. 2018a).

Another tactic of classifying HDACi is based on HDACs halted by these marvellous molecules. Certain HDACi, such as panobinostat, vorinostat and other hydroxamates, impede more than one class of HDACs and are thus deemed as pan-HDACi (Hontecillas-Prieto et al. 2020). Such inhibitors are greatly potent and three such inhibitors have previously cleared FDA expectations (Bondarev et al. 2021). Selective inhibitors, on the other hand, target either single class isozymes or solely a unique HDAC. While the former have garnered the name class selective, the hindmost are regarded as isozyme selective inhibitors (Ononye et al. 2012). Entinostat, mocetinostat and other inhibitors serve the archetype for class selective inhibitors, while tubacin is exemplar of isozyme selective inhibitors (Butler et al. 2010; Hideshima et al. 2016). Isozyme selective HDACi are preferred for neuronal complications (Thomas 2009).

1.4   DISQUIETUDE OF SYNTHETIC HDACi

Synthetic HDACi have shown very good results in various preclinical and clinical studies. Vorinostat, panobinostat and belinostat are quite successful against haematological malignancies. They have shown sanguine outcome against a wide range of cancers. Most of the synthetic inhibitors have proved many effects when they were used along with conventional anticancer therapeutics. In certain cases, they have demonstrated additive effect, while in others they have shown synergistic effect. In various solid tumours, HDACi have been reported to hamper tumour progression. These inhibitors induce differentiation, cell cycle arrest and apoptosis and autophagy in cancer models (Shao et al. 2004). HDACi exert antimetastatic effect and thus change epithelial to mesenchymal transition into mesenchymal to epithelial transition. In pancreatic cancer cells it has been found that vorinostat up-regulates the expression of E-cadherin through down-regulation of transcription factor ZEB1 (Kumagai et al. 2007; Singh et al. 2016). Pharmacological intervention with entinostat and vorinostat facilitates the efficient recognition of tumour cells by cytotoxic T cells. Exposure to aforementioned HDACi highly sensitized the breast, prostate and pancreatic cancer lines to lysis mediated by T-cells (Gameiro et al. 2016; Shanmugam,

Rakshit & Sarkar 2022). HDAC inhibitor MPT0G612 substantially decreased the viability and proliferation of colorectal cancer cells. This death was further enhanced on using autophagy inhibitor or knockdown of Atg5 (critical autophagy factor) (Chen et al. 2019). HDAC inhibitor CG200745 sensitized the gemcitabine-resistant pancreatic cancer cells to gemcitabine. Combined treatment with these inhibitors resulted in antitumour effect, which was found to be synergistic. About 50% reduction in tumour size was noticed when CG200745 was used in combinatorial fashion with gemicitabine/erlotinib under *in vivo* set up (Lee et al. 2017).

Despite these benefits, several substantial concerns are associated with the use of these (synthetic) inhibitors, which greatly dampens their clinical outcome and applicability. Off-targets effects like cardio-toxicity and diarrhoea have been reported with panobinostat. Vorinostat treatment is also associated with certain side-effects ranging from the mild to the modest, such as nausea, diarrhoea and fatigue. Apart from these modest effects, several life-threatening off-targets effects that require patient hospitalization have been encountered with vorinostat. These serious effects encompass dehydration, thrombocytopenia, squamous cell carcinoma, pulmonary embolism and severe anemia. Additionally prolongation of QTc-interval has been reported in some vorinostat-taking patients (Duvic et al. 2007). For these reasons, patients taking arrhythmic drugs are advised for daily screening if they are prescribed vorinostat. Vorinostat, being a category D drug, causes developmental abnormalities in the fetus by crossing through the placenta. Insufficient ossification of the vertebrae, axial skeleton and skull, besides the lesser birth weight of the fetus, are the developmental defects noticed with this hydroxamate HDACi in animal models (Wise, Turner & Kerr 2007; Bubna 2015). Patients having solid tumours when treated with entinostat showed fatigue, anemia and hypophosphatemia as side-effects (Connolly, Rudek & Piekarz 2017). Similarly, mocetinostat caused several side-effects, among which fatigue was seen in majority (70%), followed by nausea and diarrhoea in 69.4 and 61.1% patients respectively (Batlevi et al. 2017). Patients taking valproic acid have shown gastrointestinal disturbances, resting tremor, escalation of hepatic enzymes and mild thrombocytopenia (Bruni & Wilder 1979).

The basic concept of HAT-HDAC working, diverse classes of HDACs, HDACi and their various groups based on two different classifications has been deeply explained. Additionally, light has been shed on certain promising effects of synthetic HDACi by citing examples of well-known HDACi. Following these concepts, the flip side of synthetic inhibitors has been scrutinized to make their off-target effects conspicuous. In the following chapter, a deeper and broader overview of natural HDACi will be provided and, most notably, their natural sources will be debated in detail.

REFERENCES

Bannister, A. J. & Kouzarides, T. (2011). Regulation of chromatin by histone modifications. *Cell research 21*: 381–395.

Batlevi, C. L., Crump, M., Andreadis, C., Rizzieri, D., Assouline, S. E., Fox, S., van der Jagt, R. H. C., Copeland, A., Potvin, D., Chao, R. & Younes, A. (2017). A phase 2 study of mocetinostat, a histone deacetylase inhibitor, in relapsed or refractory lymphoma. *Br J Haematol 178*: 434–441.

Bondarev, A. D., Attwood, M. M., Jonsson, J., Chubarev, V. N., Tarasov, V. V. & Schiöth, H. B. (2021). Recent developments of HDAC inhibitors: Emerging indications and novel molecules. *British Journal of Clinical Pharmacology 87*: 4577–4597.

Bruni, J. & Wilder, B. J. (1979). Valproic acid. Review of a new antiepileptic drug. *Arch Neurol 36*: 393–398.

Bubna, A. K. (2015). Vorinostat – An Overview. *Indian Journal of Dermatology 60*: 419–419.

Butler, K. V., Kalin, J., Brochier, C., Vistoli, G., Langley, B. & Kozikowski, A. P. (2010). Rational design and simple chemistry yield a superior, neuroprotective HDAC6 inhibitor, tubastatin A. *J Am Chem Soc 132*: 10842–10846.

Chen, M.-C., Lin, Y.-C., Liao, Y.-H., Liou, J.-P. & Chen, C.-H. (2019). MPT0G612, a novel HDAC6 inhibitor, induces apoptosis and suppresses IFN-γ-induced programmed death-ligand 1 in human colorectal carcinoma cells. *Cancers (Basel) 11*: 1617.

Connolly, R. M., Rudek, M. A. & Piekarz, R. (2017). Entinostat: a promising treatment option for patients with advanced breast cancer. *Future oncology (London, England) 13*: 1137–1148.

Duvic, M., Talpur, R., Ni, X., Zhang, C., Hazarika, P., Kelly, C., Chiao, J. H., Reilly, J. F., Ricker, J. L., Richon, V. M. & Frankel, S. R. (2007). Phase 2 trial of oral vorinostat (suberoylanilide hydroxamic acid, SAHA) for refractory cutaneous T-cell lymphoma (CTCL). *Blood 109*: 31–39.

Gameiro, S. R., Malamas, A. S., Tsang, K. Y., Ferrone, S. & Hodge, J. W. (2016). Inhibitors of histone deacetylase 1 reverse the immune evasion phenotype to enhance T-cell mediated lysis of prostate and breast carcinoma cells. *Oncotarget 7*: 7390–7402.

Ganai, S. A. (2018). Histone deacetylase inhibitors modulating non-epigenetic players: the novel mechanism for small molecule based therapeutic intervention. *Curr Drug Targets 19*: 593–601.

Ganai, S. A. (2020a). Overview of epigenetic signatures and their regulation by epigenetic modification enzymes. *Histone Deacetylase Inhibitors in Combinatorial Anticancer Therapy*. Singapore, Springer Singapore: 1–33.

Ganai, S. A. (2020b). Strong involvement of classical histone deacetylases and mechanistically distinct sirtuins in bellicose cancers. *Histone Deacetylase Inhibitors in Combinatorial Anticancer Therapy*. Singapore, Springer Singapore: 75–95.

Ganai, S. A. (2020c). Summa of erasers of histone acetylation with special emphasis on classical histone deacetylases (HDACs). *Histone Deacetylase Inhibitors in Combinatorial Anticancer Therapy*. Singapore, Springer Singapore: 67–74.

Garcia-Manero, G., Abaza, Y., Takahashi, K., Medeiros, B. C., Arellano, M., Khaled, S. K., Patnaik, M., Odenike, O., Sayar, H., Tummala, M., Patel, P., Maness-Harris, L., Stuart, R., Traer, E., Karamlou, K., Yacoub, A., Ghalie, R., Giorgino, R. & Atallah, E. (2019). Pracinostat plus azacitidine in older patients with newly diagnosed acute myeloid leukemia: results of a phase 2 study. *Blood advances 3*: 508–518.

Gibney, E. R. & Nolan, C. M. (2010). Epigenetics and gene expression. *Heredity 105*: 4–13.

Glozak, M. A. & Seto, E. (2007). Histone deacetylases and cancer. *Oncogene 26*: 5420–5432.

Greer, E. L. & Shi, Y. (2012). Histone methylation: a dynamic mark in health, disease and inheritance. *Nat Rev Genet 13*: 343–357.

Hai, R., He, L., Shu, G. & Yin, G. (2021). Characterization of histone deacetylase mechanisms in cancer development. *Front Oncol*: 11.

Hideshima, T., Qi, J., Paranal, R. M., Tang, W., Greenberg, E., West, N., Colling, M. E., Estiu, G., Mazitschek, R., Perry, J. A., Ohguchi, H., Cottini, F., Mimura, N., Görgün, G., Tai, Y.-T., Richardson, P. G., Carrasco, R. D., Wiest, O., Schreiber, S. L., Anderson, K. C.

& Bradner, J. E. (2016). Discovery of selective small-molecule HDAC6 inhibitor for overcoming proteasome inhibitor resistance in multiple myeloma. *Proceedings of the National Academy of Sciences 113*: 13162–13167.

Hontecillas-Prieto, L., Flores-Campos, R., Silver, A., de Álava, E., Hajji, N. & García-Domínguez, D. J. (2020). Synergistic enhancement of cancer therapy using HDAC inhibitors: opportunity for clinical trials. *Frontiers in Genetics*: 11.

Kumagai, T., Wakimoto, N., Yin, D., Gery, S., Kawamata, N., Takai, N., Komatsu, N., Chumakov, A., Imai, Y. & Koeffler, H. P. (2007). Histone deacetylase inhibitor, suberoylanilide hydroxamic acid (Vorinostat, SAHA) profoundly inhibits the growth of human pancreatic cancer cells. *Int J Cancer 121*: 656–665.

Lee, H. S., Park, S. B., Kim, S. A., Kwon, S. K., Cha, H., Lee, D. Y., Ro, S., Cho, J. M. & Song, S. Y. (2017). A novel HDAC inhibitor, CG200745, inhibits pancreatic cancer cell growth and overcomes gemcitabine resistance. *Scientific Reports 7*: 41615.

Li, G., Tian, Y. & Zhu, W.-G. (2020). The roles of histone deacetylases and their inhibitors in cancer therapy. *Front Cell Dev Biol* 8:576946.

Licandro, S. A., Crippa, L., Pomarico, R., Perego, R., Fossati, G., Leoni, F. & Steinkühler, C. (2021). The pan HDAC inhibitor Givinostat improves muscle function and histological parameters in two Duchenne muscular dystrophy murine models expressing different haplotypes of the LTBP4 gene. *Skeletal Muscle 11*: 19.

Lin, H. Y., Chen, C. S., Lin, S. P., Weng, J. R. & Chen, C. S. (2006). Targeting histone deacetylase in cancer therapy. *Med Res Rev 26*: 397–413.

Minucci, S. & Pelicci, P. G. (2006). Histone deacetylase inhibitors and the promise of epigenetic (and more) treatments for cancer. *Nat Rev Cancer 6*: 38–51.

Mwakwari, S. C., Patil, V., Guerrant, W. & Oyelere, A. K. (2010). Macrocyclic histone deacetylase inhibitors. *Curr Top Med Chem 10*: 1423–1440.

Noureen, N., Rashid, H. & Kalsoom, S. (2010). Identification of type-specific anticancer histone deacetylase inhibitors: road to success. *Cancer Chemother Pharmacol 66*: 625–633.

Ononye, S. N., Heyst, M. v., Falcone, E. M., Anderson, A. C. & Wright, D. L. (2012). Toward isozyme-selective inhibitors of histone deacetylase as therapeutic agents for the treatment of cancer. *Pharmaceutical Patent Analyst 1*: 207–221.

Park, S.-Y. & Kim, J.-S. (2020). A short guide to histone deacetylases including recent progress on class II enzymes. *Exp Mol Med 52*: 204–212.

Rajan, A., Shi, H. & Xue, B. (2018). Class I and II histone deacetylase inhibitors differentially regulate thermogenic gene expression in brown adipocytes. *Scientific Reports 8*: 13072.

Richters, A. & Koehler, A. (2017). Epigenetic modulation using small molecules - targeting histone acetyltransferases in disease. *Curr Med Chem 24:* 4121–4150

Ropero, S. & Esteller, M. (2007). The role of histone deacetylases (HDACs) in human cancer. *Molecular Oncology 1*: 19–25.

Rossaert, E., Pollari, E., Jaspers, T., Van Helleputte, L., Jarpe, M., Van Damme, P., De Bock, K., Moisse, M. & Van Den Bosch, L. (2019). Restoration of histone acetylation ameliorates disease and metabolic abnormalities in a FUS mouse model. *Acta Neuropathologica Communications 7*: 107.

Saha, R. N. & Pahan, K. (2006). HATs and HDACs in neurodegeneration: a tale of disconcerted acetylation homeostasis. *Cell Death Differ 13*: 539–550.

Salvador, L. A. & Luesch, H. (2012). Discovery and mechanism of natural products as modulators of histone acetylation. *Curr Drug Targets 13*: 1029–1047.

Shanmugam, G., Rakshit, S. & Sarkar, K. (2022). HDAC inhibitors: targets for tumor therapy, immune modulation and lung diseases. *Translational Oncology 16*: 101312.

Shao, Y., Gao, Z., Marks, P. A. & Jiang, X. (2004). Apoptotic and autophagic cell death induced by histone deacetylase inhibitors. *Proceedings of the National Academy of Sciences of the United States of America 101*: 18030–18035.

Sheikh, B. N. (2014). Crafting the brain – role of histone acetyltransferases in neural development and disease. *Cell and Tissue Research 356*: 553–573.

Singh, A., Patel, V. K., Jain, D. K., Patel, P. & Rajak, H. (2016). Panobinostat as pan-deacetylase inhibitor for the treatment of pancreatic cancer: recent progress and future prospects. *Oncology and Therapy 4*: 73–89.

Sun, Y., Li, J., Xu, Z., Xu, J., Shi, M. & Liu, P. (2019). Chidamide, a novel histone deacetylase inhibitor, inhibits multiple myeloma cells proliferation through succinate dehydrogenase subunit A. *American Journal of Cancer Research 9*: 574–584.

Thomas, E. A. (2009). Focal nature of neurological disorders necessitates isotype-selective histone deacetylase (HDAC) inhibitors. *Molecular Neurobiology 40*: 33–45.

Wise, L. D., Turner, K. J. & Kerr, J. S. (2007). Assessment of developmental toxicity of vorinostat, a histone deacetylase inhibitor, in Sprague-Dawley rats and Dutch Belted rabbits. *Birth Defects Res B Dev Reprod Toxicol 80*: 57–68.

Yang, X. J. & Seto, E. (2007). HATs and HDACs: from structure, function and regulation to novel strategies for therapy and prevention. *Oncogene 26*: 5310–5318.

Zhang, L., Zhang, J., Jiang, Q., Zhang, L. & Song, W. (2018a). Zinc binding groups for histone deacetylase inhibitors. *Journal of Enzyme Inhibition and Medicinal Chemistry 33*: 714–721.

Zhang, Q., Wang, T., Geng, C., Zhang, Y., Zhang, J., Ning, Z. & Jiang, Z. (2018b). Exploratory clinical study of chidamide, an oral subtype-selective histone deacetylase inhibitor, in combination with exemestane in hormone receptor-positive advanced breast cancer. *Chin J Cancer Res 30*: 605–612.

2 Punctilious Overview of Stratification of Natural Histone Deacetylase Inhibitors and Their Different Provenances

Contorted acetylation homeostasis-mediated disruption of cellular integrity due to unusual histone deacetylase expression culminates in several disorders, and cancer is no exception. Resetting this distorted balance through interruption of histone deacetylases (HDACs) seems to be a propitious approach to abrogate haematological and solid malignancies. Molecules occluding the deacetylase activity of HDACs, namely HDACi, reinstate the normalcy and thus vanquish disease phenotypes. Serious and profound off-target effects observed with the usage of synthetic HDACi have switched the focus of the scientific community towards inhibitors derived from natural sources, especially from plants and marine sources. However, prior to the discussion of the anticancer property of natural HDACi, it is highly advisable to provide a conspectus of these inhibitors, their sources and their different groups.

Natural histone deacetylase inhibitors (HDACi) encompass molecules with HDAC inhibitory potential, from bacteria, fungi, algae, marine organisms and last, but no way least, from bryophytes, pteridophytes, gymnosperms and angiosperms. These natural inhibitors are classified mainly on a structural basis. The classification of these inhibitors will be rigorously discussed in this chapter in the context of recent and reliable information available about this topic.

2.1 THOROUGH CLASSIFICATION OF NATURAL HDAC INHIBITORS

Large numbers of HDACi have been derived from various sources, ranging from bacteria, fungi, marine organisms and plant sources. Termed natural HDACi, these have shown good results in various preclinical models. Natural HDACi may be cyclic peptides such as chlamydocin, azumamide E, Apicidin and HC-toxin; depsipeptides including largazole, romidepsin and spiruchostatin A; hydroxamates

DOI: 10.1201/9781003294863-2

9

encompassing TSA and amamistatin; flavonoids such as silibinin, quercetin, luteolin, genistein, apigenin, chrysin and daidzein; isothiocyanates covering sulforaphane and phenethyl isothiocyanate (PEITC); organosulfur compounds (OSCs) embracing bis (4-hydroxybenzyl) sulfide, allyl mercaptan (AM) and diallyl disulfide (DADS); short chain fatty acids enclosing propionate, butyrate and valeric acid; stilbenes, like resveratrol and piceattanol; polyketides such as depudecin and epicocconigrones A; bromotyrosine derivatives, including psammaplin A, B, E and F; coumarin derivatives confining dihydrocoumarin, prenylated isoflavones encircling pomiferin; proteins including MCP30; purine alkaloids, such as caffeine (trimethylxanthine).

2.2 SOURCES OF NATURAL HDAC INHIBITORS

As mentioned above, natural HDACi have been isolated from a variety of sources. Among these sources, bacteria, fungi, marine creatures and plants are more prominent. Only one natural HDAC inhibitor has been approved by the US FDA, and good numbers are undergoing different stages of clinical trials. Sources of natural HDACi taken from trustworthy sources are discussed below.

2.2.1 SOURCE OF CYCLIC TETRAPEPTIDE HDACI

Chlamydocin, a cyclic tetrapeptide that strongly inhibits Class I HDACs at nanomolar concentration, has been isolated from the fungus *Diheterospora chlamydosporia* (De Schepper et al. 2003). Apicidin, another cyclic tetrapeptide, has been isolated from a well-known fungus *Fusarium semitectum*, formerly known as *Fusarium pallidoroseum* (Jin et al. 2010). This fungal metabolite manifests antiparasitic activity by obstructing apicomplexan histone deacetylase at low nanomolar range (Darkin-Rattray et al. 1996). Apicidin also targets human HDACs of Class I and has shown considerable cytotoxic effect against various cancer lines (Ahn 2018). Azumamide E, another member of this group isolated from *Mycale izuensis*, a sponge, also targets Class I HDACs (Maulucci et al. 2007). *Helminthosporium carbonum* (HC)-toxin is the cyclic tetrapeptide HDACi produced by fungus pathogenic on maize, namely *Cochliobolus carbonum*. This toxin has the ability to inhibit HDACs of mammals, insects and plants. Non-ribosomal synthetase is the critical enzyme responsible for the biosynthesis of this toxin (Walton 2006). In intrahepatic cholangiocarcinoma cells, HC-toxin diminished the protein levels of HDAC1 (Zhou et al. 2016).

2.2.2 NATURAL HYDROXAMATE HDACI SOURCE

One of the earliest HDACi, trichostatin A (TSA), which serves as an antifungal antibiotic and potent pan-histone deacetylase inhibitor, has been derived from *Streptomyces hygroscopicus* (Yoshida et al. 1990). Like Vorinostat, TSA works even at lower concentrations and has proved effective against a multitude of malignancies (Vigushin et al. 2001). From *Nocardia asteroides*, an actinomycete, amamistatin A and B have been isolated (Figure 2.1). Both these molecules have shown good results on multiple cancer lines at micromolar concentration (Fennell, Möllmann & Miller 2008).

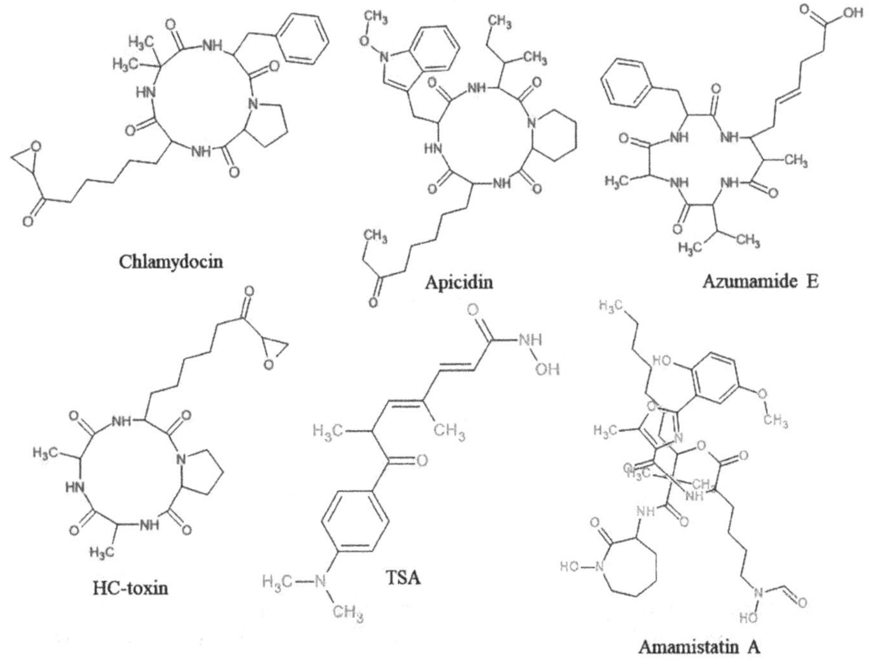

Chlamydocin

Apicidin

Azumamide E

HC-toxin

TSA

Amamistatin A

FIGURE 2.1 Detailed structure of four cyclic tetrapeptide HDACi (chlamydocin to HC-toxin) and two hydroxamate HDACi (TSA and amamistatin A). All the structures were prepared by using a popular tool ACD/ChemSketch (Freeware).

2.2.3 ORIGIN OF DEPSIPEPTIDE HDACI

Largazole, a cyclic depsipeptide, has been derived from a cyanobacterium belonging to genus *Symploca*. Thus, it is a natural product having marine source and possesses a novel chemical structure. This depsipeptide shows strong inhibitory activity against isozymes of Class I (Hong & Luesch 2012). Romidepsin is the only natural source derived HDAC inhibitor that has gained FDA approval for treating specific haematological malignancies (CTCL and PTCL). The source of this inhibitor is *Chromobacterium violaceum* (Saraiva et al. 2018). Studies have shown that romidepsin (FK228) is more effective against HDAC1 and HDAC2 over Class II HDACs, namely HDAC4 and HDAC6 (Furumai et al. 2002). Spiruchostatin A-D were purified from *Pseudomonas* species (Figure 2.2). In these bacteria, spiruchostatins were found to have gene expression facilitating function (Masuoka et al. 2001).

2.2.4 FLAVONOID HDACI AND THEIR NATURAL SOURCES

Many flavonoids have recently emerged as promising anticancer molecules. Among these molecules, apigenin, luteolin, chrysin, quercetin and silibinin are prominent. Apigenin, a trihydroxy flavone, occurs in good quantity in fruit and vegetables. Flowers and buds of *Hypericum perforatum* are an important source of apigenin.

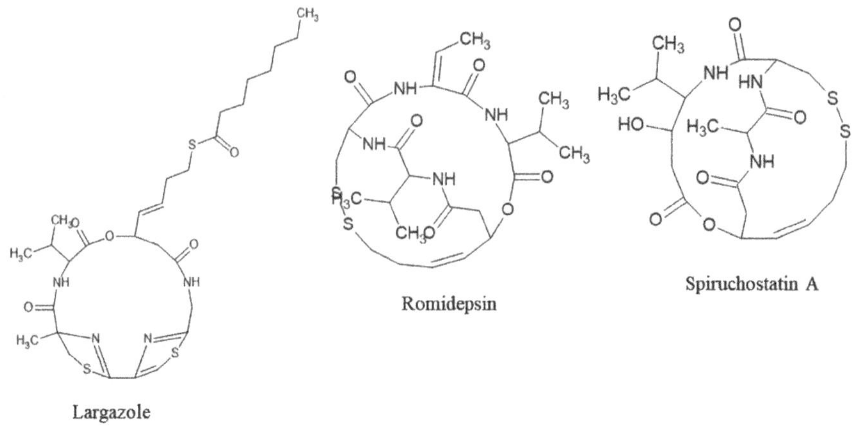

FIGURE 2.2 Different members of depsipeptide group of HDAi. Like cyclic tetrapeptides they have complex structure. Romidepsin is the sole HDAC inhibitor among natural HDACi that has cleared FDA approval.

Parsley, oranges, onion, grapes and chamomile are other sources of this natural HDAC inhibitor. This inhibitor has been found to inhibit HDAC1 and HDAC2 in prostate cancer models (Ganai 2017; Yan et al. 2017). Luteolin exists in the form of glycoside in plants and its free form is setfree during the process of absorption in the human body. A certain portion of this flavonoid is transformed to glucuronides during its passage through the mucosa of intestines. This flavone inhibits many HDACs of Class I and Class II. The maximum inhibitory potential of luteolin has been found against HDAC2 and HDAC3 (Imran et al. 2019). Parsley, peppers, onion leaves, celery, apple skin, cabbages and chrysanthemum flowers are prominent sources of luteolin. 1,035.0 mg of luteolin/kg dry weight has been quantified from bird chilli, while 74.5mg/kg from broccoli and 33.0 mg/kg from green chilli (Miean & Mohamed 2001; Ganai et al. 2021b).

Chrysin occurs in propolis, *Pelargonium crispum* and *Oroxylum indicum*. This dihydroxy flavone has showing encouraging results against neoplasms. Methoxy-chrysin and oroxylin-A, the two chrysin derivatives, have been isolated from the dry stem bark of *O. indicum*. While chrysin inhibits HDAC2 indirectly its inhibition on HDAC8 is both direct and indirect (Pal-Bhadra et al. 2012; Ganai, Sheikh & Baba 2021a). Silibinin, another inhibitor of this family, has been extracted from *Silybum marianum L* (milk thistle). It has been proved to down-regulate the expression of two Class I HDACs (HDAC1 and HDAC2). Chemically, silibinin is flavonolignan that has demonstrated optimistic effect in prostate carcinoma (Anestopoulos et al. 2016). Quercetin falls within the flavonol subgroup of flavonoids and has 3, 3', 4', 5, 7-pentahydroxyflvanone as its IUPAC name. Quercetin occurs in Brassica vegetables, grapes, shallots, tomatoes, berries, apples, seeds, flowers, leaves and nuts. Apart from these sources, quercetin has many other sources, such as *Sambucus canadensis*, *Ginkgo biloba* and *Hypericum perforatum* (Häkkinen et al. 1999). A greater amount

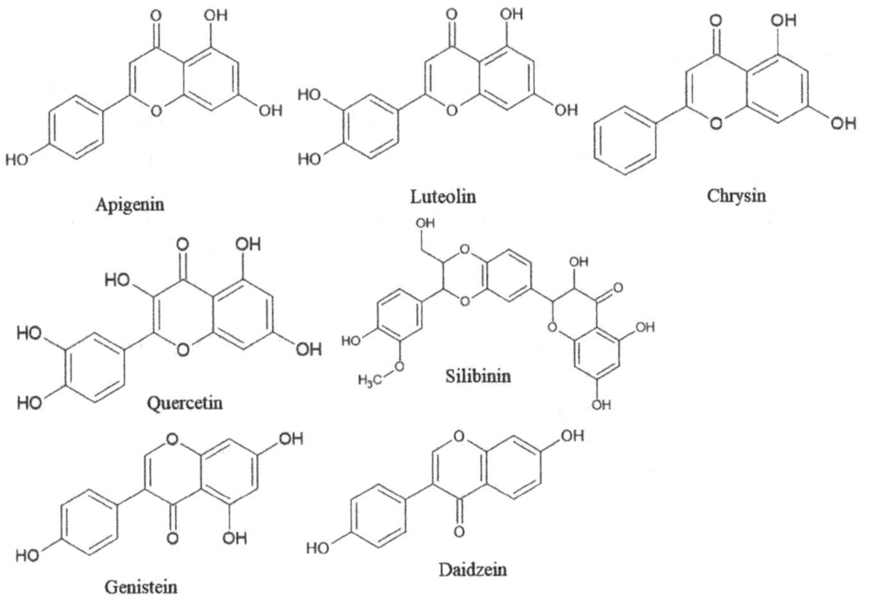

FIGURE 2.3   Structural overview of different flavonoid HDACi.

The majority of natural HDACi fall within the zone of flavonoids. Apigenin, luteolin and chrysin are the most well studied among the flavonoid group of HDACi.

of quercetin is present in that portion of red onion that is spaced in the vicinity of root (Smith et al. 2003; Wiczkowski et al. 2008). Tomatoes that are organically raised have 79% more quercetin content over those developed by using chemical methods (Mitchell et al. 2007). Evidence suggests the absorption of quercetin in the upper section of the small intestines, and no reports are available suggesting its absorption in the stomach (Crespy et al. 1999). However, it is well established that both quercetin and its derivatives demonstrate stability in the environment of gastric acid (Li et al. 2016).

Genistein, an isoflavone derived from soy, has shown some sort of HDAC inhibitory activity. This crux has been derived as prolonged treatment with this inhibitor caused hyperacetylation of core histone H3 in breast cancer cells (Jawaid et al. 2010). This isoflavone has been observed to reduce the expression of HDAC6 through hyperacetylation of Hsp90 (Yamaki et al. 2011). Daidzein, like genistein, is an isoflavone that provides protection against osteoporosis, cancer and ischemic heart disease (Figure 2.3) (Liggins et al. 2000). Its main sources are soya beans, soy flours and other legumes. Certain plants such as *Pueraria mirifica* and *Pueraria lobata* are the other sources of daidzein (Liggins et al. 2002; Francisco et al. 2018). Daidzein and other phytoestrogens were found to modulate site-specific methylation marks and histone H3 lysine 4 acetylation (H3K4ac) in breast cancer lines (Dagdemir et al. 2013).

2.2.5   IMPORTANT SOURCES OF ISOTHIOCYANATE GROUP OF HDACI

Sulforaphane is the most well-studied member of this group. Its main sources are broccoli (44–171 mg/100g dry weight) and broccoli sprouts (1153 mg/100g dry weight). It is present in foods in the form of isothiocyanate glucoraphanin. With the help of the myrosinase enzyme produced by the gut, microflora glucoraphanin is transformed to sulforaphane (Fahey et al. 2015; Ganai 2016). Sulforaphane has been found to target Class I HDACs and two members of Class II, namely HDAC4 and HDAC6 (Li, Li & Guo 2010; Abbaoui et al. 2017). The major bioactive constituents present in cruciferous vegetables are isothiocyanates. In many cruciferous vegetables, phenethyl isothiocyanate (PEITC) occurs as gluconasturtiin (Gupta et al. 2014). Apart from water cress and broccoli, which are regarded as the two of its richest sources, PEITC also occurs in radish and turnips. This inhibitor is released from its precursor gluconasturtiin through the assistance of myrosinase (Shapiro et al. 1998; Getahun & Chung 1999; Gupta et al. 2014). This inhibitor has been reported to decrease the activity of HDAC1 and HDAC2 in prostate cancer cells (Beklemisheva et al. 2006). PEITC was found to facilitate the expression of p21 through inhibition of HDACs and other player known as c-Myc (Wang et al. 2008).

2.2.6   MAIN SOURCES OF ORGANOSULFUR HDACI

Bis (4-hydroxybenzyl) sulfide, a sulphur containing HDAC inhibitor has *Pleuropterus ciliinervis* root as its source (Son et al. 2007b). This inhibitor was also isolated from *Gastrodia elata* (Chinese herbal plant) and has demonstrated excellent potency against mushroom tyrosinase (Chen et al. 2015). Bis (4-hydroxybenzyl) sulfide has shown inhibitory activity against HDACs of cancer cells (Son et al. 2007b). Sulphur possessing HDAC inhibitor allyl mercaptan (AM) has been derived from garlic and has been noticed to inhibit HDACs (Nian et al. 2008). Diallyl disulfide (diallyl disulfide), another member present in this group, has garlic as its source (Figure 2.4). This dietary agent was reported to have HDAC inhibitory effect, like sulforaphane and butyrate (Myzak, Ho & Dashwood 2006).

2.2.7   ORIGINAL SOURCES OF SHORT CHAIN FATTY ACID HDACI

One of the main short chain fatty acid HDAC inhibitor generated from dietary fibres through fermentation mediated by gut flora is butyrate. The main sources of this HDAC inhibitor are soya beans, peas, beans, nuts, cereals, fruits and whole grains. Besides, this inhibitor also occurs in cheese and butter (Schnekenburger & Diederich 2015). Leaving Class III and Class IIb HDACs, butyrate targets remaining HDACs (Davie 2003). Propionate is considered as another natural inhibitor of HDACs (Sivaprakasam et al. 2017). This inhibitor has potential to inhibit HDAC2 and HDAC8 (Silva et al. 2018). Valeric acid is another natural HDAC inhibitor that belongs to short chain fatty acids (Yuille et al. 2018). Its natural sources are *Valeriana officinalis* and *Angelica archangelica* (Goldberg & Rokem 2009). One study has reported HDACs 1–3 as targets of valeric acid (Han et al. 2020).

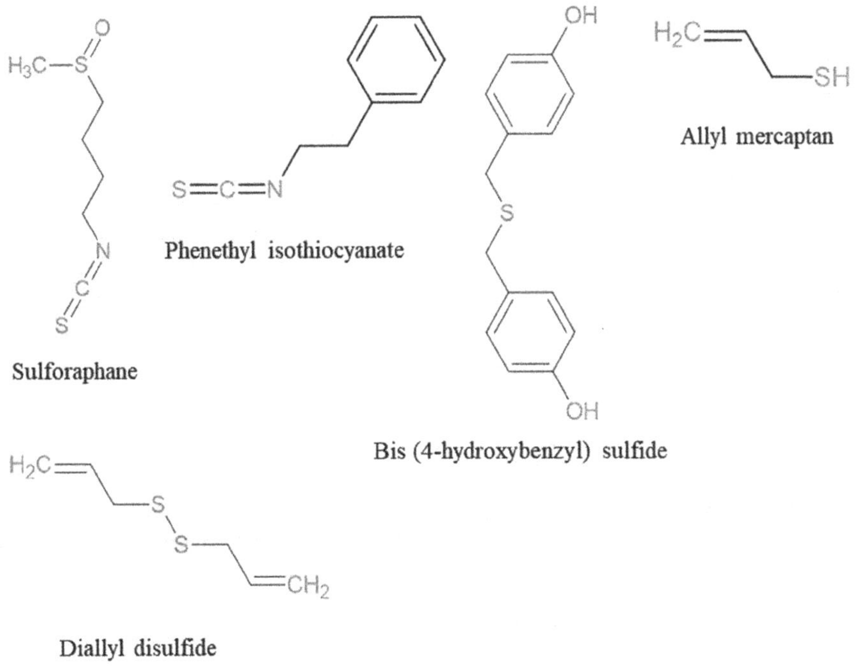

FIGURE 2.4   Isothiocyanate HDACi and their chemical structures (sulforaphane and PEITC).

Organosulfur HDACi from the third structure and onwards contain one or more sulphur atoms.

2.2.8   STILBENES AND THEIR PREMIER SOURCES

Resveratrol is an HDAC inhibitor that falls within the confines of the stilbene family. The predominant sources of resveratrol is grape skin, where it has quantified as 50–100 µg/g (Li et la. 2006). Evidence-based findings suggest that bilberries and blueberries also contain resveratrol, but its content is comparatively lesser (below 10%) as compared to levels reported for grapes (Lyons et al. 2003). Resveratrol acts as a pan-HDAC inhibitor and was found to inhibit all classical HDACs. The best concentration for this inhibition was found to be 100 µM (Venturelli et al. 2013).

Piceatannol, the polyphenolic stilbene possessing antiinflammatory activity, occurs in certain fruits and vegetables. Its concentration is more in red grapes (374 ng/g) than white grapes (43 ng/g) (Viñas et al. 2011). The seed of the passion fruit also contains high amount of this inhibitor (4.8 mg/g) (Matsui et al. 2010). Vaccinium berries are also an important source of piceatannol (Rimando et al. 2004; Kershaw & Kim 2017). This resveratrol analog inhibited various classical HDACs, including all Class I HDACs except HDAC3 and a majority of Class IIa HDACs, including HDAC5, HDAC7-9 (Hsu et al. 2016).

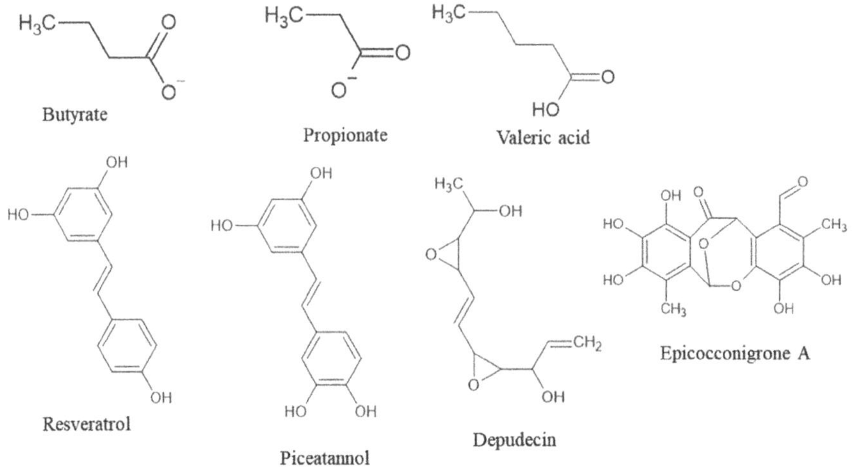

FIGURE 2.5   Natural short chain fatty acid group inhibitor structures (butyrate to valeric acid), some examples of stilbenes (resveratrol and piceatannol) and polyketide group HDACi.

2.2.9   CORE SOURCES OF POLYKETIDES

The HDAC inhibitor depudecin is the prime example of this group. It is composed of 11 carbon atoms and is a linear polyketides. Depudecin is produced by *Alternaria brassicicola*, a pathogenic fungus (Wight et al. 2009). *In vitro* study has found its inhibitory effect on HDAC1 (Kwon et al. 1998). Epicocconigrones A is one more polyketide group HDAC inhibitor from *Epicoccum nigrum*, an endophytic fungus (Figure 2.5) (El Amrani et al. 2014). Class I, HDAC11 and Class IIb are the target HDACs of this inhibitor (Losson et al. 2016).

2.2.10   BEST SOURCES OF HDACi BELONGING TO BROMO-TYROSINE DERIVATIVES

This group of HDACi covers psammaplins. Among the various psammaplins, the most widely studied bromo-tyrosine derivative is psammaplin A. Psammaplins have been retrieved from *Pseudoceratina purpurea*, a demosponge. Bisaprasin and psammaplin F are also potent HDACi that have demonstrated mild cytotoxic effect (Piña et al. 2003). Comparison of the HDAC inhibitory strength of psammaplins (A, B, E and F) revealed that psammaplin A is highly potent compared to others, and psammaplin F has lesser potency than the remaining three (Figure 2.6). Experimental evidences support psammaplin A as an inhibitor selective towards Class I HDACs (Kim, Shin & Kwon 2007).

2.2.11   NATURAL ORIGIN OF COUMARIN DERIVATIVES, PRENYLATED ISOFLAVONES AND OTHER HDACi

Coumarin derivative such as dihydrocoumarin, has shown inhibitory activity against SIR2 of yeast. This inhibitor has *Melilotus officinalis* as its source (Olaharski et al.

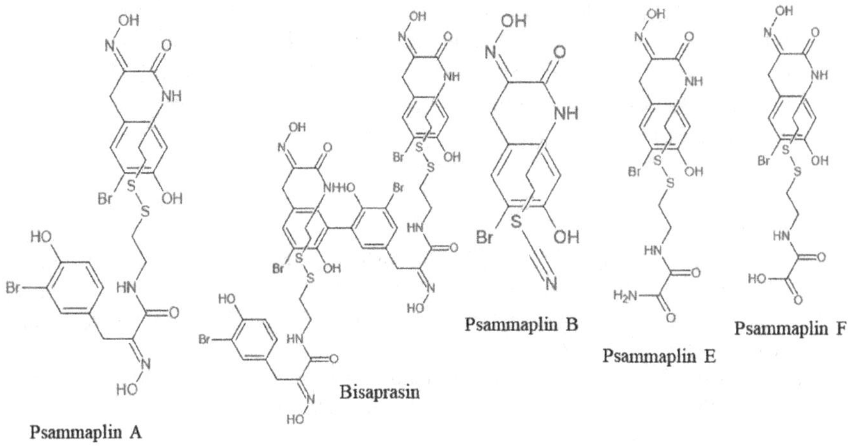

FIGURE 2.6 Structures of natural HDACi of the bromo-tyrosine derivatives group. Among these psammaplin A has demonstrated relatively potent HDAC inhibitory function.

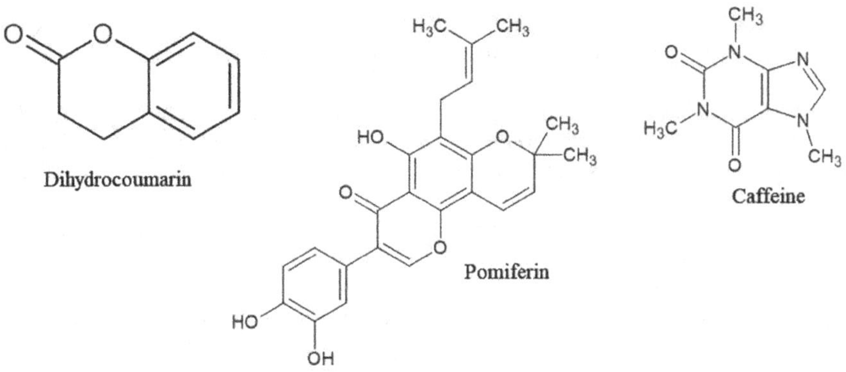

FIGURE 2.7 Chemical illustrations of dihydrocoumarin (coumarin derivative), pomiferin (prenylated isoflavone) and caffeine (purine alkaloid).

2005). Pomiferin, the prenylated isoflavone-type HDAC inhibitor has been confirmed from the fruits of *Maclura pomifera* (osage orange tree), along with osajin. Pomiferin has proved to be effective in restraining Class I HDACs (Son et al. 2007a; Seidel et al. 2012). MCP30, a protein derived from *Momordica charantia*, exhibited HDAC1 inhibitory activity and facilitated acetylation of core histones H3 and H4 in cell models of prostate cancer (Xiong et al. 2009). Caffeine, the main sources of which are coffee and tea, has proved to be HDACi (Figure 2.7). In glioma cells, caffeine substantially reduced the activity of HDAC1 and concurrently enhanced the p300 activity (Chen & Hwang 2016).

To this point, various groups of natural HDACi along with their respective archetypes, their major sources and the possible HDAC enzymes targeted by them, have been debated. From this classification, it seems that majority of natural HDACi come under the jurisdiction of the flavonoid group of HDACi. Soy, broccoli, fungi, bacteria, sponges and plants serve as the reservoirs of these promising molecules. In the next chapter, therefore, it will be quite interesting to shed light on the anticancer activity of different natural HDAC inhibitor groups.

REFERENCES

Abbaoui, B., Telu, K. H., Lucas, C. R., Thomas-Ahner, J. M., Schwartz, S. J., Clinton, S. K., Freitas, M. A. & Mortazavi, A. (2017). The impact of cruciferous vegetable isothiocyanates on histone acetylation and histone phosphorylation in bladder cancer. *Journal of Proteomics 156*: 94–103.

Ahn, M. Y. (2018). HDAC inhibitor apicidin suppresses murine oral squamous cell carcinoma cell growth in vitro and in vivo via inhibiting HDAC8 expression. *Oncology Letters 16*: 6552–6560.

Anestopoulos, I., Sfakianos, A., Franco, R., Chlichlia, K., Panayiotidis, M., Kroll, D. & Pappa, A. (2016). A novel role of silibinin as a putative epigenetic modulator in human prostate carcinoma. *Molecules 22*: 62.

Beklemisheva, A. A., Fang, Y., Feng, J., Ma, X., Dai, W. & Chiao, J. W. (2006). Epigenetic mechanism of growth inhibition induced by phenylhexyl isothiocyanate in prostate cancer cells. *Anticancer Res 26*: 1225–1230.

Chen, J.-C. & Hwang, J.-H. (2016). Effects of caffeine on cell viability and activity of histone deacetylase 1 and histone acetyltransferase in glioma cells. *Ci ji yi xue za zhi = Tzu-chi medical journal 28*: 103–108.

Chen, W. C., Tseng, T. S., Hsiao, N. W., Lin, Y. L., Wen, Z. H., Tsai, C. C., Lee, Y. C., Lin, H. H. & Tsai, K. C. (2015). Discovery of highly potent tyrosinase inhibitor, T1, with significant anti-melanogenesis ability by zebrafish in vivo assay and computational molecular modeling. *Sci Rep 5*: 7995.

Crespy, V., Morand, C., Manach, C., Besson, C., Demigne, C. & Remesy, C. (1999). Part of quercetin absorbed in the small intestine is conjugated and further secreted in the intestinal lumen. *Am J Physiol 277*: G120–126.

Dagdemir, A., Durif, J., Ngollo, M., Bignon, Y. J. & Bernard-Gallon, D. (2013). Histone lysine trimethylation or acetylation can be modulated by phytoestrogen, estrogen or anti-HDAC in breast cancer cell lines. *Epigenomics 5*: 51–63.

Darkin-Rattray, S. J., Gurnett, A. M., Myers, R. W., Dulski, P. M., Crumley, T. M., Allocco, J. J., Cannova, C., Meinke, P. T., Colletti, S. L., Bednarek, M. A., Singh, S. B., Goetz, M. A., Dombrowski, A. W., Polishook, J. D. & Schmatz, D. M. (1996). Apicidin: a novel antiprotozoal agent that inhibits parasite histone deacetylase. *Proceedings of the National Academy of Sciences of the United States of America 93*: 13143–13147.

Davie, J. R. (2003). Inhibition of histone deacetylase activity by butyrate. *J Nutr 133*: 2485S–2493S.

De Schepper, S., Bruwiere, H., Verhulst, T., Steller, U., Andries, L., Wouters, W., Janicot, M., Arts, J. & Van Heusden, J. (2003). Inhibition of histone deacetylases by chlamydocin induces apoptosis and proteasome-mediated degradation of survivin. *J Pharmacol Exp Ther 304*: 881–888.

El Amrani, M., Lai, D., Debbab, A., Aly, A. H., Siems, K., Seidel, C., Schnekenburger, M., Gaigneaux, A., Diederich, M., Feger, D., Lin, W. & Proksch, P. (2014). Protein Kinase and HDAC inhibitors from the endophytic fungus epicoccum nigrum. *Journal of natural products 77*: 49–56.

Fahey, J. W., Holtzclaw, W. D., Wehage, S. L., Wade, K. L., Stephenson, K. K. & Talalay, P. (2015). Sulforaphane bioavailability from glucoraphanin-rich broccoli: control by active endogenous myrosinase. *PLoS One 10*: e0140963–e0140963.

Fennell, K. A., Möllmann, U. & Miller, M. J. (2008). Syntheses and biological activity of amamistatin B and analogs. *J Org Chem 73*: 1018–1024.

Francisco, V., Costa, G., Neves, B. M., Cruz, M. T. & Batista, M. T. (2018). Chapter 31 – Antiinflammatory activity of polyphenols on dendritic cells. *Polyphenols: Prevention and Treatment of Human Disease (Second Edition)*. R. R. Watson, V. R. Preedy and S. Zibadi (eds). London, Academic Press: 395–415.

Furumai, R., Matsuyama, A., Kobashi, N., Lee, K. H., Nishiyama, M., Nakajima, H., Tanaka, A., Komatsu, Y., Nishino, N., Yoshida, M. & Horinouchi, S. (2002). FK228 (depsipeptide) as a natural prodrug that inhibits class I histone deacetylases. *Cancer Res 62*: 4916–4921.

Ganai, S. A. (2016). Histone deacetylase inhibitor sulforaphane: the phytochemical with vibrant activity against prostate cancer. *Biomed Pharmacother 81*: 250–257.

Ganai, S. A. (2017). Plant-derived flavone Apigenin: the small-molecule with promising activity against therapeutically resistant prostate cancer. *Biomed Pharmacother 85*: 47–56.

Ganai, S. A., Sheikh, F. A. & Baba, Z. A. (2021a). Plant flavone chrysin as an emerging histone deacetylase inhibitor for prosperous epigenetic-based anticancer therapy. *Phytotherapy Research n/a*.

Ganai, S. A., Sheikh, F. A., Baba, Z. A., Mir, M. A., Mantoo, M. A. & Yatoo, M. A. (2021b). Anticancer activity of the plant flavonoid luteolin against preclinical models of various cancers and insights on different signalling mechanisms modulated. *Phytother Res 35*: 3509–3532.

Getahun, S. M. & Chung, F. L. (1999). Conversion of glucosinolates to isothiocyanates in humans after ingestion of cooked watercress. *Cancer Epidemiol Biomarkers Prev 8*: 447–451.

Goldberg, I. & Rokem, J. S. (2009). Organic and fatty acid production, microbial. *Encyclopedia of Microbiology (Third Edition)*. M. Schaechter (ed.). Oxford, Academic Press: 421–442.

Gupta, P., Wright, S. E., Kim, S.-H. & Srivastava, S. K. (2014). Phenethyl isothiocyanate: a comprehensive review of anti-cancer mechanisms. *Biochim Biophys Acta 1846*: 405–424.

Häkkinen, S. H., Kärenlampi, S. O., Heinonen, I. M., Mykkänen, H. M. & Törrönen, A. R. (1999). Content of the flavonols quercetin, myricetin, and kaempferol in 25 edible berries. *J Agric Food Chem 47*: 2274–2279.

Han, R., Nusbaum, O., Chen, X. & Zhu, Y. (2020). Valeric acid suppresses liver cancer development by acting as a novel HDAC inhibitor. *Molecular Therapy – Oncolytics 19*: 8–18.

Hong, J. & Luesch, H. (2012). Largazole: from discovery to broad-spectrum therapy. *Nat Prod Rep 29*: 449–456.

Hsu, C.-W., Shou, D., Huang, R., Khuc, T., Dai, S., Zheng, W., Klumpp-Thomas, C. & Xia, M. (2016). Identification of HDAC inhibitors using a cell-based HDAC I/II Assay. *Journal of Biomolecular Screening 21*: 643–652.

Imran, M., Rauf, A., Abu-Izneid, T., Nadeem, M., Shariati, M. A., Khan, I. A., Imran, A., Orhan, I. E., Rizwan, M., Atif, M., Gondal, T. A. & Mubarak, M. S. (2019). Luteolin, a flavonoid, as an anticancer agent: a review. *Biomedicine & Pharmacotherapy 112*: 108612.

Jawaid, K., Crane, S. R., Nowers, J. L., Lacey, M. & Whitehead, S. A. (2010). Long-term genistein treatment of MCF-7 cells decreases acetylated histone 3 expression and alters growth responses to mitogens and histone deacetylase inhibitors. *J Steroid Biochem Mol Biol 120*: 164–171.

Jin, J. M., Lee, S., Lee, J., Baek, S. R., Kim, J. C., Yun, S. H., Park, S. Y., Kang, S. & Lee, Y. W. (2010). Functional characterization and manipulation of the apicidin biosynthetic pathway in Fusarium semitectum. *Mol Microbiol 76*: 456–466.

Kershaw, J. & Kim, K.-H. (2017). The therapeutic potential of piceatannol, a natural stilbene, in metabolic diseases: a review. *Journal of Medicinal Food 20*: 427–438.

Kim, D. H., Shin, J. & Kwon, H. J. (2007). Psammaplin A is a natural prodrug that inhibits class I histone deacetylase. *Exp Mol Med 39*: 47–55.

Kwon, H. J., Owa, T., Hassig, C. A., Shimada, J. & Schreiber, S. L. (1998). Depudecin induces morphological reversion of transformed fibroblasts via the inhibition of histone deacetylase. *Proceedings of the National Academy of Sciences of the United States of America 95*: 3356–3361.

Li, X., Wu, B., Wang, L. & Li, S. (2006). Extractable amounts of trans-resveratrol in seed and berry skin in Vitis evaluated at the germplasm level. *J Agric Food Chem 54*: 8804–8811.

Li, Y., Li, X. & Guo, B. (2010). Chemopreventive agent 3, 3'-diindolylmethane selectively induces proteasomal degradation of class I histone deacetylases. *Cancer Res 70*: 646–654.

Li, Y., Yao, J., Han, C., Yang, J., Chaudhry, M. T., Wang, S., Liu, H. & Yin, Y. (2016). Quercetin, inflammation and immunity. *Nutrients 8*: 167.

Liggins, J., Bluck, L. J., Runswick, S., Atkinson, C., Coward, W. A. & Bingham, S. A. (2000). Daidzein and genistein content of fruits and nuts. *J Nutr Biochem 11*: 326–331.

Liggins, J., Mulligan, A., Runswick, S. & Bingham, S. A. (2002). Daidzein and genistein content of cereals. *Eur J Clin Nutr 56*: 961–966.

Losson, H., Schnekenburger, M., Dicato, M. & Diederich, M. (2016). Natural compound histone deacetylase inhibitors (HDACi): synergy with inflammatory signaling pathway modulators and clinical applications in cancer. *Molecules 21*: 1608.

Lyons, M. M., Yu, C., Toma, R. B., Cho, S. Y., Reiboldt, W., Lee, J. & van Breemen, R. B. (2003). Resveratrol in raw and baked blueberries and bilberries. *J Agric Food Chem 51*: 5867–5870.

Masuoka, Y., Nagai, A., Shin-ya, K., Furihata, K., Nagai, K., Suzuki, K.-i., Hayakawa, Y. & Seto, H. (2001). Spiruchostatins A and B, novel gene expression-enhancing substances produced by Pseudomonas sp. *Tetrahedron Letters 42*: 41–44.

Matsui, Y., Sugiyama, K., Kamei, M., Takahashi, T., Suzuki, T., Katagata, Y. & Ito, T. (2010). Extract of passion fruit (Passiflora edulis) seed containing high amounts of piceatannol inhibits melanogenesis and promotes collagen synthesis. *J Agric Food Chem 58*: 11112–11118.

Maulucci, N., Chini, M. G., Di Micco, S., Izzo, I., Cafaro, E., Russo, A., Gallinari, P., Paolini, C., Nardi, M. C., Casapullo, A., Riccio, R., Bifulco, G. & De Riccardis, F. (2007). Molecular insights into azumamide E histone deacetylases inhibitory activity. *Journal of the American Chemical Society 129*: 3007–3012.

Miean, K. H. & Mohamed, S. (2001). Flavonoid (myricetin, quercetin, kaempferol, luteolin, and apigenin) content of edible tropical plants. *J Agric Food Chem 49*: 3106–3112.

Mitchell, A. E., Hong, Y. J., Koh, E., Barrett, D. M., Bryant, D. E., Denison, R. F. & Kaffka, S. (2007). Ten-year comparison of the influence of organic and conventional crop management practices on the content of flavonoids in tomatoes. *J Agric Food Chem 55*: 6154–6159.

Myzak, M. C., Ho, E. & Dashwood, R. H. (2006). Dietary agents as histone deacetylase inhibitors. *Mol Carcinog 45*: 443–446.

Nian, H., Delage, B., Pinto, J. T. & Dashwood, R. H. (2008). Allyl mercaptan, a garlic-derived organosulfur compound, inhibits histone deacetylase and enhances Sp3 binding on the P21WAF1 promoter. *Carcinogenesis 29*: 1816–1824.

Olaharski, A. J., Rine, J., Marshall, B. L., Babiarz, J., Zhang, L., Verdin, E. & Smith, M. T. (2005). The flavoring agent dihydrocoumarin reverses epigenetic silencing and inhibits sirtuin deacetylases. *PLoS Genet 1*: e77.

Pal-Bhadra, M., Ramaiah, M. J., Reddy, T. L., Krishnan, A., Pushpavalli, S., Babu, K. S., Tiwari, A. K., Rao, J. M., Yadav, J. S. & Bhadra, U. (2012). Plant HDAC inhibitor chrysin arrest cell growth and induce p21WAF1by altering chromatin of STAT response element in A375 cells. *BMC cancer 12*: 180.

Piña, I. C., Gautschi, J. T., Wang, G.-Y.-S., Sanders, M. L., Schmitz, F. J., France, D., Cornell-Kennon, S., Sambucetti, L. C., Remiszewski, S. W., Perez, L. B., Bair, K. W. & Crews, P. (2003). Psammaplins from the sponge pseudoceratina purpurea: inhibition of both histone deacetylase and DNA methyltransferase. *J Org Chem 68*: 3866–3873.

Rimando, A. M., Kalt, W., Magee, J. B., Dewey, J. & Ballington, J. R. (2004). Resveratrol, pterostilbene, and piceatannol in vaccinium berries. *J Agric Food Chem 52*: 4713–4719.

Saraiva, R. G., Huitt-Roehl, C. R., Tripathi, A., Cheng, Y.-Q., Bosch, J., Townsend, C. A. & Dimopoulos, G. (2018). Chromobacterium spp. mediate their anti-Plasmodium activity through secretion of histone deacetylase inhibitor romidepsin. *Scientific Reports 8*: 6176.

Schnekenburger, M. & Diederich, M. (2015). Chapter 18 Nutritional epigenetic regulators in the field of cancer: new avenues for chemopreventive approaches. *Epigenetic Cancer Therapy*. S. G. Gray (ed.). Boston, Academic Press: 393–425.

Seidel, C., Schnekenburger, M., Dicato, M. & Diederich, M. (2012). Histone deacetylase modulators provided by Mother Nature. *Genes & Nutrition 7*: 357–367.

Shapiro, T. A., Fahey, J. W., Wade, K. L., Stephenson, K. K. & Talalay, P. (1998). Human metabolism and excretion of cancer chemoprotective glucosinolates and isothiocyanates of cruciferous vegetables. *Cancer Epidemiol Biomarkers Prev 7*: 1091–1100.

Silva, L. G., Ferguson, B. S., Avila, A. S. & Faciola, A. P. (2018). Sodium propionate and sodium butyrate effects on histone deacetylase (HDAC) activity, histone acetylation, and inflammatory gene expression in bovine mammary epithelial cells. *Journal of Animal Science 96*: 5244–5252.

Sivaprakasam, S., Bhutia, Y. D., Ramachandran, S. & Ganapathy, V. (2017). Cell-surface and nuclear receptors in the colon as targets for bacterial metabolites and its relevance to colon health. *Nutrients 9*: 856.

Smith, C., Lombard, K. A., Peffley, E. B. & Liu, W. (2003). Genetic analysis of quercetin in onion (Allium allium cepa L.) 'Lady Raider'. *Texas Journal of Agriculture and Natural Resources 16*: 24–28.

Son, I., Chung, I.-M., Lee, S., Yang, H.-D. & Moon, H.-I. (2007a). Pomiferin, histone deacetylase inhibitor isolated from the fruits of Maclura pomifera. *Bioorganic & Medicinal Chemistry Letters 17*: 4753–4755.

Son, I. H., Lee, S. I., Yang, H. D. & Moon, H. I. (2007b). Bis(4-hydroxybenzyl)sulfide: a sulfur compound inhibitor of histone deacetylase isolated from root extract of Pleuropterus ciliinervis. *Molecules 12*: 815–820.

Venturelli, S., Berger, A., Böcker, A., Busch, C., Weiland, T., Noor, S., Leischner, C., Schleicher, S., Mayer, M., Weiss, T. S., Bischoff, S. C., Lauer, U. M. & Bitzer, M. (2013). Resveratrol as a pan-HDAC inhibitor alters the acetylation status of histone [corrected] proteins in human-derived hepatoblastoma cells. *PLoS One 8*: e73097–e73097.

Vigushin, D. M., Ali, S., Pace, P. E., Mirsaidi, N., Ito, K., Adcock, I. & Coombes, R. C. (2001). Trichostatin A is a histone deacetylase inhibitor with potent antitumor activity against breast cancer in vivo. *Clin Cancer Res 7*: 971–976.

Viñas, P., Martínez-Castillo, N., Campillo, N. & Hernández-Córdoba, M. (2011). Directly suspended droplet microextraction with in injection-port derivatization coupled to gas chromatography-mass spectrometry for the analysis of polyphenols in herbal infusions, fruits and functional foods. *J Chromatogr A 1218*: 639–646.

Walton, J. D. (2006). HC-toxin. *Phytochemistry 67*: 1406–1413.

Wang, L. G., Liu, X. M., Fang, Y., Dai, W., Chiao, F. B., Puccio, G. M., Feng, J., Liu, D. & Chiao, J. W. (2008). De-repression of the p21 promoter in prostate cancer cells by an isothiocyanate via inhibition of HDACs and c-Myc. *Int J Oncol 33*: 375–380.

Wiczkowski, W., Romaszko, J., Bucinski, A., Szawara-Nowak, D., Honke, J., Zielinski, H. & Piskula, M. K. (2008). Quercetin from shallots (Allium allium cepa L. var. aggregatum) is more bioavailable than its glucosides. *J Nutr 138*: 885–888.

Wight, W. D., Kim, K. H., Lawrence, C. B. & Walton, J. D. (2009). Biosynthesis and role in virulence of the histone deacetylase inhibitor depudecin from *Alternaria brassicicola*. *Mol Plant Microbe Interact 22*: 1258–1267.

Xiong, S., Yu, K., Liu, X.-H., Yin, L., Kirschenbaum, A., Yao, S., Narla, G., Difeo, A., Wu, J., Yuan, Y., Ho, S.-M., Lam, Y. & Levine, A. (2009). Ribosome-inactivating proteins isolated from dietary bitter melon induce apoptosis and inhibit histone deacetylase-1 selectively in premalignant and malignant prostate cancer cells. Int J Cancer 125: 774–782. *International Journal of Cancer. Journal International du Cancer 125*: 774–782.

Yamaki, H., Nakajima, M., Shimotohno, K. W. & Tanaka, N. (2011). Molecular basis for the actions of Hsp90 inhibitors and cancer therapy. *J Antibiot (Tokyo) 64*: 635–644.

Yan, X., Qi, M., Li, P., Zhan, Y. & Shao, H. (2017). Apigenin in cancer therapy: anti-cancer effects and mechanisms of action. *Cell & Bioscience 7*: 50–50.

Yoshida, M., Kijima, M., Akita, M. & Beppu, T. (1990). Potent and specific inhibition of mammalian histone deacetylase both in vivo and in vitro by trichostatin A. *J Biol Chem 265*: 17174–17179.

Yuille, S., Reichardt, N., Panda, S., Dunbar, H. & Mulder, I. E. (2018). Human gut bacteria as potent class I histone deacetylase inhibitors in vitro through production of butyric acid and valeric acid. *PLoS One 13*: e0201073.

Zhou, W., Chen, X., He, K., Xiao, J., Duan, X., Huang, R., Xia, Z., He, J., Zhang, J. & Xiang, G. (2016). Histone deacetylase inhibitor screening identifies HC toxin as the most effective in intrahepatic cholangiocarcinoma cells. *Oncol Rep 35*: 2535–2542.

3 Natural Cyclic Tetrapeptide Histone Deacetylase Inhibitors and Their Optimistic Role in Anticancer Therapy

It has been proved beyond doubt that histone deacetylase inhibitors (HDACi) are the best among the emerging epigenetic therapeutics against cancer. These inhibitors, by way of tuning the deviated acetylation homeostasis, restore correct transcriptional events critical for balanced cellular functioning. Cyclic tetrapeptide inhibitors of natural origin come under the banner of macrocyclic HDACi. These inhibitors consist of complex cap regions that form various interactions with HDAC surface residues. As these residues vary from HDAC to HDAC, thus alterations in cap region of these inhibitors can be utilized for customizing isozyme selective cyclic tetrapeptide inhibitors. Chlamydocin, apicidin and azumamide E are the three prominent examples of natural inhibitors of the cyclic-tetrapeptide group that have shown encouraging clinical outcomes against a broad range of malignancies.

3.1 ANTICANCER EFFECT OF CHLAMYDOCIN

Chlamydocin exhibits stronger antiproliferative effect and inhibits HDACs at nanomolar concentration. In ovarian cancer cells, this inhibitor enhanced the acetylation profile of core histones and enhanced the p21 (cip1/waf1) expression. Chlamydocin induced apoptosis in these cells through activation of caspase-3 and subsequent cleavage of p21 (cip1/waf1). Following this, the levels of survivin which is well known for selective expression in tumour cells, get alleviated as chlamydocin triggers its proteasomal degradation (De Schepper et al. 2003). Compared to actinomycin D, colchicine and other anticancer agents, chlamydocin manifested potent cytostatic activity against mouse mastocytoma cells (P-*815*) (Stähelin & Trippmacher 1974).

DOI: 10.1201/9781003294863-3

3.2 PROMISING ANTICANCER PROPERTY OF APICIDIN

The anticancer activity of apicidin has been tested on a variety of cancer cells and animal models. Human endometrial cancer Ishikawa cells were subcutaneously administered in nude mice, following which the apoptosis and cell proliferation were measured after the intervention with this inhibitor. A substantial increase in the acetylation of histone H3 was noticed in Ishikawa cells on apicidin treatment under *in vitro* conditions. Apicidin treatment caused a marked decline in the expression of HDAC4 and HDAC3. Apicidin hampered tumour growth of the xenograft model, mitigated the expression of vascular endothelial growth factor (VEGF) and proliferative cell nuclear antigen (PCNA). From these findings it is evident that the antitumour effect of apicidin is strongly mediated through the suppression of a couple of HDACs (HDAC3 and HDAC4) (Ahn et al. 2010).

One more study evaluated the apoptotic effect of apicidin on two cancer cell lines (endometrial and breast cancer) and one normal line. The two cancer lines showed sensitivity towards apicidin while the normal cells were quite viable after the treatment, clearly suggesting that the growth inhibitory activity of this inhibitor is selective to cancer cells. Apicidin induced apoptosis in these cells, as indicated by the loss of mitochondrial membrane potential and exposure of phosphatidylserine. The acetylation status of tails of H3 and H4 was escalated by apicidin-based therapeutic intervention (Ueda et al. 2007). The anticancer effect of apicidin has also been evaluated on human salivary mucoepidermoid carcinoma cells (YD-15). A marked inhibition in the proliferation of these cells was observed on exposure to apicidin. The induction of apoptosis was noticed, which was accompanied with extracellular signal-regulated kinase (ERK) and AKT/mTOR signalling inactivation. Apicidin treatment culminated in c-Jun NH2-terminal kinase (JNK) activation. On the other hand, autophagy was induced by this inhibitor by way of obstructing AKT/mTOR signaling. Insulin-like growth factor 1 receptor (IGF-1R) acts upstream to MAPK and AKT/mTOR and, as such, apicidin-induced effects in the above-mentioned cells are ascribed to down-regulation of this growth factor (Ahn et al. 2015).

Another study explored the antitumour potential of apicidin using the murine oral squamous cell carcinoma (AT-84) cell model. Apicidin restrained growth of these cells and lowered expression of Class I HDAC8. Not only apoptosis but also induction of autophagy was seen on exposure to apicidin. The subcutaneous injection of these cells in the C3H mouse model was used for the created *in vivo* set up. Apicidin administration was found to decrease tumour growth by 46% compared to respective control. High expression of HDAC8 was noticed in tumour cells, which got substantially reduced in the apicidin-treated group as compared to the group receiving vehicle. Taken together, apicidin-triggered cytotoxic effect in AT-84 cells and rodent model is mediated through down-regulation of HDAC8 (Ahn 2018). The effect of apicidin derived from secondary metabolites of Fusarium was studied on GLC-82 (non-small cell lung cancer) - cells. Strong inhibitory action was seen in these cells after treatment with apicidin. The proliferation of these cells was blocked by this inhibitor and induction of apoptosis was noticed. The apoptosis was confirmed to be mediated through the mitochondrial pathway as a decline in mitochondrial membrane potential, cytosolic release of cytochrome c, activation of caspase-9 and 3 besides poly-ADP-ribose polymerase cleavage was noticeable in treated cells (Zhang et al. 2017).

In pancreatic cancer cells (Panc-1 and Capan-1) cytotoxicity of apicidin was studied in time and dose-dependent fashion. Both short- and long-term exposure of apicidin was given to these cells prior to cytotoxic assays. Substantial cytotoxicity and loss of viability of cells was observed with prolonged and continuous apicidin treatment. Antiproliferative effect was noticed in the aforesaid lines before the onset of cytotoxic effect (Bauden, Tassidis & Ansari 2015). Further, in mouse Neuro-2a neuroblastoma cells, apicidin incited apoptotic death through activation of multiple caspases, including caspase-12 and 9. Induction of endoplasmic reticulum stress-related proteins such as CHOP, production of reactive oxygen species and loss of mitochondrial membrane potential was seen following apicidin intervention. These effects were rescued by N-acetyl cysteine, an antioxidant suggesting that generation of ROS is critical for apicidin-induced cell death (Choi et al. 2012).

The underlying molecular mechanism by which apicidin modulates invasion and migration of ovarian cancer (SKOV-3) cells was examined. From the study it was found that this inhibitor prevented cell migration and invasion by alleviating the expression of HDAC4, a Class II isozyme. Apicidin by way of down-regulating the HDAC4 suppressed its binding to RECK promoters Sp1 binding elements. Inhibition of cell migration was accompanied with up-regulation of RECK and matrix metalloproteinase-2 (MMP-2) down-regulation. Under conditions of *in vivo*, transplanted SKOV-3 cell growth was curbed by apicidin through HDAC4 and MMP-2 down-regulation (Ahn et al. 2012). Hypoxia-inducible factor prolyl hydroxylase 2 (HIF-PH2) has cross-talk with angiogenesis. Enhanced angiogenesis is triggered by diminished expression of HIF-PH2. A variety of cancer cells including SiHa, C33A, HeLa and CaSki were employed to study the impact of apicidin on PH2 enzyme. Pharmacological intervention with apicidin enhanced the mRNA and protein levels of this enzyme in all cells except SiHa. Further scrutinizing through bisulfite sequencing revealed that apicidin causes demethylation of CpG islands spaced in the exon (first one) of PH2 gene. As reduced PH2 expression facilitates the progression of tumour thus apicidin having the tendency to up-regulate its expression may serve as a promising anticancer agent (Durczak & Jagodzinski 2010).

In T cells, phosphorylation of phospholipase C gamma-1 (PLCgamma1) is crucial for intracellular signal transduction. The effect of apicidin on this phospholipase was studied on Hut-78 T lymphoma cells. In these cells, it was observed that apicidin reduced the expression of PLCgamma1 both at both levels (mRNA and protein). Apicidin also shortened the half-life of mRNA of PLCgamma1 significantly (Debicki & Jagodzinski 2009). Oncogenic proteins of human papillomavirus (HPV), namely E6 and E7, have the ability of inactivating Rb and p53, culminating in malignancy. In cervical cancer (SiHa) cells, apicidin caused the marked decline in HPV16-E6 and -E7 mRNA and protein levels. The stability of transcripts of HPV16-E6 and -E7 was substantially reduced by apicidin. While the stability of E6 was shortened from 5h to 2h, the stability of E7 was reduced from 6h to 3h (Luczak & Jagodzinski 2008).

The growth of breast cancer cells is restrained by apicidin. A study was undertaken to delineate the effect of apicidin on the proliferation of breast cancer lines.

Apicidin-treated cells demonstrated a remarkable increase in the acetylation profile of core histones H3 and H4. While the expression of both ERalpha and ERbeta was lowered in MCF-7 cells, no change in the expression of the latter was observed in MDA-MB-231 cells. Substantial induction in the expression of p21Waf1 and p27Kip1 occurred at 300 nM concentration of apicidin. A significant decline in the expression of cell cycle regulatory molecules, including cyclin E/CDK 2 and cyclin D1/CDK 4, was evident only from MCF-7 cells. The bax/bcl-2 ratio was substantially elevated by this cyclic tetrapeptide in MCF-7 cells. Thus, apicidin impedes the proliferation of estrogen receptor positive MCF-7 cells by alleviating the expression of proteins regulating cell cycle and through the induction of apoptosis (Im et al. 2008). DNA methyltransferase (DNMT)-1 dysregulation results in cellular transformation and its blockade yields antitumour outcome. The abnormal expression of this DNA methyltransferase occurs in HeLa cells. The HDAC inhibitor apicidin provoked downregulation of DNMT1 in these cells. Selective suppression of the DNMT1 expression was observed in HeLa cells. Although global H3 and H4 acetylation was recorded intervention with apicidin but localized deacetylation of these histones was noticed at the binding site of E2F. Si-RNA mediated knockdown of DNMT1 induced apoptosis in HeLa cells, suggesting this enzyme as a promising target to overcome cervical cancer (You et al. 2008).

Expressing oncogenic Ras in cancer cells makes them susceptible to the intervention of HDACi. The antiproliferative mechanism induced by apicidin was compared in non-transformed epithelial (MCF10A) cells and H-ras-transformed human breast epithelial (MCF10A-ras). In both these lines, apicidin elevated the acetylation status of H3 and H4. While attenuation of cyclin E and CDK2 expression occurred in MCF10A cells, declined cyclin D1 and cyclin E levels were observed in MCF10A-ras cells on exposure of apicidin. Certain changes like enhanced levels of CDK inhibitors, p27Kip1 and p21WAF1/Cip1 were evident from both these lines. Importantly, following apicidin treatment the hyperphosphorylated form of Rb protein were depleted in MCF10A-ras cells subsequent to exposure of apicidin. In MCF10A-ras cells, p53 up-regulation and ERK activation was quite evident after the defined treatment. Up-regulated p53 in turn facilitated Bax expression, resulting in the caspase-9 and caspase-6 activation, culminating in apoptosis of these cells. Moreover, ERK1/2 phosphorylation levels were markedly elevated in MCF10A-ras cells after HDAC inhibitor exposure (Park et al. 2008). The influence of apicidin on angiogenesis and cancer invasion has been studied using v-ras-transformed mouse fibroblast NIH3T3 cells. Morphological alteration and induction of different degree of acetylated forms of H4 were seen in these cells upon apicidin use. The invasion of human melanoma (A2058) cells and v-ras-NIH3T3 cells was dramatically suppressed by this promising HDACi (Kim et al. 2004). Another study examined the ability of apicidin to induce apoptosis in HL60 cells and the molecular mechanism being involved. The viability of these cells was reduced in dose-dependent fashion by this strong molecule. Morphological changes in nucleus, DNA fragmentation and formation of apoptotic bodies happened on apicidin use. The activation of procaspase-3, poly (ADP-ribose) polymerase cleavage was elicited in HL60 by apicidin (Kwon et al. 2002).

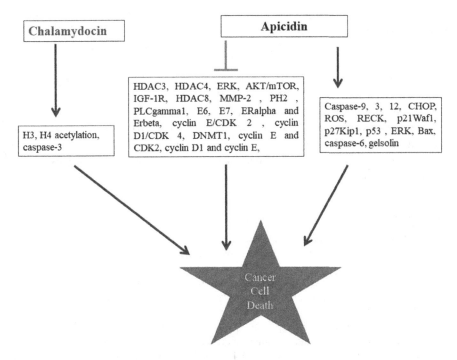

FIGURE 3.1 Conspectus of diverse molecular players modulated by chlamydocin and apicidin for inducing cytotoxic effect and inhibiting proliferation in various cancer cells. Among the epigenetic players inhibited by these inhibitors are some specific HDACs (HDAC3, 4 and 8) and DNA methyltransferase (DNMT)-1.

Apicidin exerted a broad spectrum antiproliferative effect on a variety of cancer lines to a different extent. Its antiproliferative effect on HeLa cells was associated with changes in morphology, G1 phase arrest, H4 hyperacetylation. It is known that p21WAF1/Cip1 controls cell cycle and gelsolin manages morphology. Induction of p21WAF1/Cip1 and gelsolin was the outcome of apicidin treatment. It is clear from above that apicidin-induced H4 hyperacetylation results in induction of the above-noted proteins, thereby causing cell cycle arrest and changes in cell morphology (Figure 3.1) (Han et al. 2000).

Azumamide E derived from *Mycale izuensis*, a marine sponge in leukemia cells (K562), has shown potent HDAC inhibitory effect (Nakao et al. 2006). This inhibitor showed selective inhibition towards Class I HDACs, especially against isozyme HDAC1, 2 and HDAC3. While azumamide E showed an IC_{50} of 0.05 against HDAC1, this value for HDAC2 and HDAC3 was found to be 0.1 and 0.01 µM respectively (Maulucci et al. 2007). Another study revealed that azumamide E and C are strong inhibitors of Class IIb HDAC10 and Class IV HDAC11 (Villadsen at al. 2013). The antiangiogenic activity of azumamides isolated from marine sponge was tested on the vascular organization model in which mouse-induced pluripotent stem cells were used. Azumamide proved comparatively more effective in inhibiting angiogenesis than azumamide A. Inhibition was obtained at 1.9 μM in the case of the former and at 19 μM in the latter case (Deng et al. 2020).

The importance of the cyclic tetrapeptide group of macrocyclic HDACi in cancer therapy has been explained. Different molecular mechanisms by which these inhibitors induce cell-cycle arrest and curb proliferation of distinct cancer cells have also been taken into account. The impact of these inhibitors on acetylation profile of two core histones H3 and H4 in cancer models has been thoroughly put together. Insights about various HDACs and DNMTs targeted by cyclic tetrapeptides for exerting anticancer effect in multiple cancer cell lines have been provided. In the next chapter, the anticancer potential of natural hydroxamates and depsipeptides will be extensively discussed.

REFERENCES

Ahn, M. Y. (2018). HDAC inhibitor apicidin suppresses murine oral squamous cell carcinoma cell growth in vitro and in vivo via inhibiting HDAC8 expression. *Oncology Letters* *16*: 6552–6560.

Ahn, M.-Y., Ahn, J.-W., Kim, H.-S., Lee, J. & Yoon, J.-H. (2015). Apicidin inhibits cell growth by downregulating IGF-1R in salivary mucoepidermoid carcinoma cells. *Oncol Rep* *33*: 1899–1907.

Ahn, M. Y., Chung, H. Y., Choi, W. S., Lee, B. M., Yoon, S. & Kim, H. S. (2010). Anti-tumor effect of apicidin on Ishikawa human endometrial cancer cells both in vitro and in vivo by blocking histone deacetylase 3 and 4. *Int J Oncol 36*: 125–131.

Ahn, M. Y., Kang, D. O., Na, Y. J., Yoon, S., Choi, W. S., Kang, K. W., Chung, H. Y., Jung, J. H., Min do, S. & Kim, H. S. (2012). Histone deacetylase inhibitor, apicidin, inhibits human ovarian cancer cell migration via class II histone deacetylase 4 silencing. *Cancer Lett 325*: 189–199.

Bauden, M., Tassidis, H. & Ansari, D. (2015). In vitro cytotoxicity evaluation of HDAC inhibitor apicidin in pancreatic carcinoma cells subsequent time and dose dependent treatment. *Toxicology Letters 236*: 8–15.

Choi, J. H., Lee, J. Y., Choi, A. Y., Hwang, K. Y., Choe, W., Yoon, K. S., Ha, J., Yeo, E. J. & Kang, I. (2012). Apicidin induces endoplasmic reticulum stress- and mitochondrial dysfunction-associated apoptosis via phospholipase Cγ1- and Ca(2+)-dependent pathway in mouse Neuro-2a neuroblastoma cells. *Apoptosis 17*: 1340–1358.

De Schepper, S., Bruwiere, H., Verhulst, T., Steller, U., Andries, L., Wouters, W., Janicot, M., Arts, J. & Van Heusden, J. (2003). Inhibition of histone deacetylases by chlamydocin induces apoptosis and proteasome-mediated degradation of survivin. *J Pharmacol Exp Ther 304*: 881–888.

Debicki, S. & Jagodzinski, P. P. (2009). Apicidin decreases phospholipase C gamma-1 transcript and protein in Hut-78 T lymphoma cells. *Biomed Pharmacother 63*: 543–547.

Deng, B., Luo, Q., Halim, A., Liu, Q., Zhang, B. & Song, G. (2020). The antiangiogenesis role of histone deacetylase inhibitors: their potential application to tumor therapy and tissue repair. *DNA Cell Biol 39*: 167–176.

Durczak, M. & Jagodzinski, P. P. (2010). Apicidin upregulates PHD2 prolyl hydroxylase gene expression in cervical cancer cells. *Anticancer Drugs 21*: 619–624.

Han, J. W., Ahn, S. H., Park, S. H., Wang, S. Y., Bae, G. U., Seo, D. W., Kwon, H. K., Hong, S., Lee, H. Y., Lee, Y. W. & Lee, H. W. (2000). Apicidin, a histone deacetylase inhibitor, inhibits proliferation of tumor cells via induction of p21WAF1/Cip1 and gelsolin. *Cancer Res 60*: 6068–6074.

Im, J. Y., Park, H., Kang, K. W., Choi, W. S. & Kim, H. S. (2008). Modulation of cell cycles and apoptosis by apicidin in estrogen receptor (ER)-positive and-negative human breast cancer cells. *Chem Biol Interact 172*: 235–244.

Kim, S. H., Ahn, S., Han, J. W., Lee, H. W., Lee, H. Y., Lee, Y. W., Kim, M. R., Kim, K. W., Kim, W. B. & Hong, S. (2004). Apicidin is a histone deacetylase inhibitor with antiinvasive and anti-angiogenic potentials. *Biochem Biophys Res Commun 315*: 964–970.

Kwon, S. H., Ahn, S. H., Kim, Y. K., Bae, G. U., Yoon, J. W., Hong, S., Lee, H. Y., Lee, Y. W., Lee, H. W. & Han, J. W. (2002). Apicidin, a histone deacetylase inhibitor, induces apoptosis and Fas/Fas ligand expression in human acute promyelocytic leukemia cells. *J Biol Chem 277*: 2073–2080.

Luczak, M. W. & Jagodzinski, P. P. (2008). Apicidin down-regulates human papillomavirus type 16 E6 and E7 transcripts and proteins in SiHa cervical cancer cells. *Cancer Lett 272*: 53–60.

Maulucci, N., Chini, M. G., Di Micco, S., Izzo, I., Cafaro, E., Russo, A., Gallinari, P., Paolini, C., Nardi, M. C., Casapullo, A., Riccio, R., Bifulco, G. & De Riccardis, F. (2007). Molecular insights into azumamide E histone deacetylases inhibitory activity. *Journal of the American Chemical Society 129*: 3007–3012.

Nakao, Y., Yoshida, S., Matsunaga, S., Shindoh, N., Terada, Y., Nagai, K., Yamashita, J. K., Ganesan, A., van Soest, R. W. M. & Fusetani, N. (2006). Azumamides A–E: histone deacetylase inhibitory cyclic tetrapeptides from the marine sponge mycale izuensis. *Angewandte Chemie International Edition 45*: 7553–7557.

Park, H., Im, J. Y., Kim, J., Choi, W. S. & Kim, H. S. (2008). Effects of apicidin, a histone deacetylase inhibitor, on the regulation of apoptosis in H-ras-transformed breast epithelial cells. *Int J Mol Med 21*: 325–333.

Stähelin, H. & Trippmacher, A. (1974). Cytostatic activity of chlamydocin, a rapidly inactivated cyclic tetrapeptide. *European Journal of Cancer (1965) 10*: 801–808.

Ueda, T., Takai, N., Nishida, M., Nasu, K. & Narahara, H. (2007). Apicidin, a novel histone deacetylase inhibitor, has profound anti-growth activity in human endometrial and ovarian cancer cells. *Int J Mol Med 19*: 301–308.

Villadsen, J. S., Stephansen, H. M., Maolanon, A. R., Harris, P. & Olsen, C. A. (2013). Total synthesis and full histone deacetylase inhibitory profiling of azumamides A-E as well as β^2- epi-azumamide E and β^3-epi-azumamide E. *J Med Chem 56*: 6512–6520.

You, J. S., Kang, J. K., Lee, E. K., Lee, J. C., Lee, S. H., Jeon, Y. J., Koh, D. H., Ahn, S. H., Seo, D. W., Lee, H. Y., Cho, E. J. & Han, J. W. (2008). Histone deacetylase inhibitor apicidin downregulates DNA methyltransferase 1 expression and induces repressive histone modifications via recruitment of corepressor complex to promoter region in human cervix cancer cells. *Oncogene 27*: 1376–1386.

Zhang, J., Lai, Z., Huang, W., Ling, H., Lin, M., Tang, S., Liu, Y. & Tao, Y. (2017). Apicidin inhibited proliferation and invasion and induced apoptosis via mitochondrial pathway in non-small cell lung cancer GLC-82 cells. *Anticancer Agents Med Chem 17*: 1374–1382.

4 Anticancer Potential of Natural Hydroxamates and Depsipeptides Against Different Disease Models

Most of the FDA-approved HDAC inhibitors (HDACi) come under synthetic hydroxamates. Inhibitors falling within this group show comparably higher potency than other HDAC inhibitory groups. Hydroxamates target classical HDACs of more than one class and, as such, come under pan-HDAC inhibitors. These inhibitors work well against haematological malignancies, but are not equally potent against solid tumours. Depsipeptides come under the roof of macrocyclic HDACi and, because of their unique chemical structure, demonstrate different therapeutic functions. As the context of this chapter is natural hydroxamate HDACi and depsipeptides, I will discuss trichostatin A (TSA) and amamistatin under the former heading, largazole, romidepsin and spiruchostatin under the depsipeptides.

4.1 HYDROXAMATE HDACi AND THEIR ANTINEOPLASTIC EFFECT

From some decades, many drugs have emerged that have the potential to reactivate silenced genes by inhibiting all or some classical HDACs. TSA is one such inhibitor that derepresses epigenetically-suppressed tumour-controlling genes. The antiproliferative activity of TSA was evaluated against breast cancer lines and its antitumour potential was tested on an *in vivo* tumour model. Under both *in vitro* and *in vivo* conditions TSA exerted strong antitumour effect dose dependently. No detectable toxicity was observed at 5mg/kg of TSA administration subcutaneously (Vigushin et al. 2001). TSA was also examined on two breast cancer lines (MCF-7 and MDA-MB-231) and one non-tumorigenic epithelial cell line (MCF-10A). In both cancer lines, cell viability and proliferation was restrained by TSA but, at the same time, no effect was observed on the normal line. The induction of apoptosis in breast cancer cells was accompanied with G2-M arrest, which was triggered by enhanced ROS production due to TSA (Sun et al. 2014). TSA was also tested against the MCF-7TN-R breast carcinoma cell line, which is resistant to apoptosis. The outcome of the study was that TSA manifests anticancer activity even against apoptosis-resistant cell lines. This effect of TSA has been ascribed to substantial modulation of various

DOI: 10.1201/9781003294863-4

miRNAs in this cell line. Out of the 32 modulated miRNAs, 22 were found to be up-regulated whereas ten underwent down-regulation (Rhodes et al. 2012).

Evidence suggests that TSA is effective even against pancreatic cancer cells. This inhibitor elicited apoptosis in these cells through depolarization of mitochondrial membrane in addition to cytochrome c release and subsequent caspase-3 activation. Among the two cell lines, TSA was able to increase the Bim (proapoptotic), decline Mcl-1 (antiapoptotic), increase ROS generation and eventual autophagy in PaCA44 cells (Gilardini Montani et al. 2017). Evaluation of TSA was done on three pancreatic cancer lines, AsPC-1, BxPC-3 and CAPAN-1. Among the three lines TSA proved more effective against BxPC-3 cells, followed by AsPC-1 and CAPAN-1 respectively. Their response to TSA correlated with enhanced H3 acetylation. Cells possessing a wild version of KRAS (BxPC-3 cells) displayed potential inhibition of ERK1/2 phosphorylation and AKT on pharmacological intervention of TSA, while those with oncogenic KRAS (CAPAN-1 and AsPC-1) showed either no or only moderate effect on the aforesaid molecules on exposure to TSA. Wild-type KRAS containing a cell line among the concerned lines exhibited the most prominent TSA-induced MAP kinase p38 activation. TSA-triggered antiproliferative effect in BxPC-3 cells was substantially lowered through p38 inhibition by SB202190. A marked increase in bax transcript was noticeable in TSA-exposed BxPC-3 cells. Apart from this, substantially elevated p21^{Waf1} levels were recorded in BxPC-3 and AsPC-1 lines following TSA application (Emonds, Fitzner & Jaster 2010).

Cell proliferation and apoptosis in many cancer cells are monitored by TSA. Its effect on death and growth of cervical cancer cells has been studied in the context of glutathione (GSH) and ROS. Exposure of HeLa cells to TSA resulted in time- and dose-dependent growth inhibition. Apoptosis was seen in the treated cells as several indications, like caspase-3 activation, mitochondrial membrane potential loss were evident. Additionally, the effects of TSA were strengthened through usage of siRNA against Bcl-2. Further, increased ROS levels and reduction in GSH levels were associated with TSA intervention. The effects of TSA on HeLa cells were rescued by antioxidant N-acetyl cysteine. Conclusively, TSA-induced growth inhibition in cervical cancer cells was incited by elevated ROS and was mediated by Bcl-2 (You & Park 2013). The impact of TSA was further studied on CaSki and HeLa, the two cervical cancer lines. In both lines, TSA substantially reduced viability and increased the rate of apoptosis in dose-dependent fashion. This treatment was accompanied with significantly increased beclin1 and LC3 II/I ratio. On the other hand, p62 levels were sharply declined by this natural hydroxamate. In addition to this, TRPV6 and PRMT5 levels underwent suppression, while p-JNK and STC1 levels escalated after the incubation of these cells with TSA. Thus, inhibition of proliferation of cervical cancer cells and eventual induction of apoptotic death and autophagy on TSA application occurs via regulation of the PRMT5/STC1/TRPV6/JNK cascade (Liu et al. 2021).

Endoplasmic reticulum stress may be one of the possibilities by which TSA exerts anticancer effect. It has also been established that the function of endoplasmic reticulum following stress is regulated by p53 (Jeong et al. 2014). In wild-type human colon cancer (HCT116)- cells, endoplasmic reticulum stress was induced by TSA. This was evident from the substantial elevation of GRP94 and GRP78, the markers indicating endoplasmic reticulum stress. Contrary to WT HCT116 cells, the apoptosis rate declined and cell viability elevated in HCT116 TP53(-/-) cells after TSA addition. This gives an indication that TSA-invoked endoplasmic reticulum stress is reliant on the p53-dependent process in HCT116 cells (Dai et al. 2019). The viability of cisplatin-resistant and cisplatin-sensitive human ovarian cancer lines was reduced by TSA. This effect has been attributed to the ability of this inhibitor to enhance caspase-9 activity, LC3-II expression and p21 expression. The increased activity of caspase-3 has cross-talk with apoptosis progression, while the increased expression of LC3-II is a hallmark of apoptosis. The elevated expression of p53 noticed on TSA use induces cell cycle arrest. Taurine transporter up-regulation and activity has been linked to cisplatin resistance. Treatment with TSA reduces the activity and expression of this protein to values typical for cisplatin-sensitive cells. Thus, one of the aspects of TSA-induced death in cisplatin-resistant cells is alleviating the activity of the taurine transporter (Lambert, Nielsen & Stürup 2020).

Ovarian tumours come under the umbrella of fatal gynaecological cancers and currently differentiation-inducing therapies are tested for circumventing it (Reid, Permuth & Sellers 2017). In ovarian cancer cells (HO8910), TSA exposure invoked differentiation. This TSA-induced differentiation was specified by particular changes in cell morphology, enhanced expression of FOXA2 (differentiation marker), diminished expression of SOX2 and OCT4 (pluripotency markers), inhibition of cell proliferation and arrest of cells at the G1 phase. While increased p21Cip1 (protein-inhibiting cell cycle) expression was noticed after use of TSA, cyclin D1 (regulating cell cycle) levels were lowered. TSA-invoked differentiation was further intensified when epidermal growth factor signalling cascade was obstructed with inhibitors specific for this pathway (Shao et al. 2017). From the *Nocardia* strain a couple of natural products, including amamistatin A and B, have been isolated. These amamistatins have demonstrated growth-inhibitory property against multiple human tumour cell models. Amamistatin B proved to be more effective against the breast cancer (MCF-7)-line as compared to the prostate cancer (PC-3)-line. While IC_{50} of 0.12–0.20 μM was estimated against the former cell line, an IC_{50} of 8–13 μM was quantified for the latter (Fennell, Möllmann & Miller 2008).

4.2   ANTICANCER PROPERTY OF NATURAL DEPSIPEPTIDE HDACi

Like hydroxamates, the depsipeptide group of inhibitors has shown promising results in preclinical and advanced studies. Growth inhibitory effects were seen on the treatment of HCT116 colon cancer cells with largazole. In the xenograft mouse model developed through the administration of human HCT116 cells, largazole blocked HDACs of tumour tissue. This was followed by histone hyperacetylation, hampering of tumour growth and subsequent induction of apoptosis. This effect seemed to be mediated by blockade of AKT signalling due to down-regulation of insulin receptor substrate 1 and alleviation of EGFR (Liu et al. 2010).

Many biological processes having a significant role in tumour onset, progression and chemoresistance are controlled by the RNA-binding protein Musashi-2 (MSI2). This protein is overrepresented in non-small cell lung cancer (NSCLC) and reduces the clinical outcome of patients. Through a structure-based drug-designing approach, namely molecular docking, largazole was predicted to bind with MSI2. Largazole intervention lowered the expression of MSI2 both at transcript and protein level. This resulted in the inhibition of the underlying mTOR signalling cascade. In CML and NSCLC cells, largazole obstructed proliferation and invoked apoptosis. These findings establish MSI2 as a strikingly therapeutically relevant target to overcome CML and NSCLC (Wang et al. 2020).

Largazole has strong inhibitory activity against Class I HDACs and, as such, is Class I selective HDAC inhibitor. In preclinical models of solid tumours, this inhibitor has demonstrated anticancer activity. It has proved effective against a cancer, namely glioblastoma multiforme. This inhibitor elevates the expression of genes meant for neuroprotection such as Pax6 and Bdnf. This suggests that largazole is feasible for treating neurodegenerative disorders as well as brain cancer. Up-regulation of Pax6 induced by largazole has a substantial contribution in neurogenesis, cognitive ability and neuronal plasticity. Enhanced expression of Pax6 restrains proliferation as well as invasion of glioblastoma multiforme. This suggests that the anticancer agent largazole has the potential to get repurposed for neurological maladies (Al-Awadhi et al. 2020). The epigenetic properties of the genome are altered by histone deacetylase enzymes. By way of blocking the activity of these enzymes, HDACi induce differentiation, cell cycle arrest, apoptosis and inhibit migration. The effect of largazole has been evaluated on breast cancer lines and other metastatic lines under both hypoxia and normoxia conditions. This inhibitor evinced good result against estrogen receptor positive breast cancers, hampering metastasis of triple-negative breast cancer cells.

Cell cycle aberrations are implicated in lung cancer and, therefore, inhibitors of the cell cycle may prove an effective therapeutic tactic. Largazole, the macrocyclic depsipeptide, strongly inhibits the clonogenic activity and proliferation of lung cancer cells. However, no such effect has been observed on normal epithelial cells. This inhibitor induced G1 arrest that was accompanied with increased p21, the cyclin-dependent kinase inhibitor. The binding affinity between E2F1-HDAC1 was escalated by largazole, and it was found that this inhibitor induces degradation of E2F1 by proteasomes. This represses the function of E2F1 in lung cancer cells while leaving it unchanged in normal cells. On the whole, largazole functions as a cell cycle inhibitor that targets E2F1, due to which it may prove as a promising antineoplastic molecule (Wu et al. 2013).

Romidepsin, as mentioned before, has achieved FDA approval for CTCL treatment. Epigenetic alterations result in malignant transformation. HDACi tune the deviated epigenetic signatures and thus hamper cancer onset and its subsequent progression (Bertino & Otterson 2011). The advanced stages of CTCL when treated with bexarotene or other standard chemotherapeutics show relapse. The FDA approved romidepsin for CTCL based on a pair of huge phase II studies (Prince & Dickinson 2012). Romidepsin has the potential to incite cell cycle arrest and cytotoxic effect in cancer models. Its ability to induce cell cycle arrest and apoptosis are reliant on its ability to induce hyperacetylation of histones. The pharmacological effects of romidepsin have been studied on the lung cancer (A549) line. A significant decline in the cyclin B1, Cdc2/Cdk-1 and phosphorylated retinoblastoma protein (pRb), but an increase in p21 expression, was evident on treating these cells with romidepsin. At concentrations ranging from 25 to 50 nM, romidepsin induced apoptotic death in aforesaid cells. The conclusion derived from these findings is that romidepsin inhibits the proliferation of human lung cancer cells through modulation of cell cycle controlling proteins (Vinodhkumar, Song & Devaki 2008).

Studies have been performed for exploring the actual mechanism through which romidepsin induces therapeutic effect in malignant T cells. Certain cell lines, including PEER, SUPT1 (T-cell lines) and one derived from a T-cell lymphoma victim, were subjected to romidepsin. Following this, the impact of romidepsin on critical pathways related to oncogenic signalling was scrutinized. Romidepsin proved more effective against patient J cells, followed by SUPT1 and PEER cells. An IC_{50} of 7.0 nM was obtained against J cells, 7.9 nM for SUPT1 and 10.8 nM for PEER cells. The potent obstruction of demethylases, histone deacetylases, enhanced ROS production and reduced mitochondrial membrane potential were noticed. Certain signalling pathways encompassing stress-activated protein kinase/c-Jun N-terminal kinase were provoked, while pathways, such as β-catenin and phosphatidylinositol 3-kinase/AKT/mammalian target of rapamycin (PI3K/AKT/mTOR), were blocked. Romidepsin caused hypomethylation of the SFRP1 gene promoter, thereby enhancing its expression. SFRP1 serves as the inhibitor of β-catenin and thus its romidepsin-induced inhibition is mediated through SFRP1 (Valdez et al. 2015).

Romidepsin targets not only HDACs but modulates other multiple targets playing a role in cancer onset and progression, such as p53, c-MYC and HSP90 (Campas-Moya 2009). This inhibitor induces differentiation and apoptosis in acute promyelocytic leukaemia cells. Additionally, this inhibitor has shown significant beneficial effects against acute myeloid leukaemia cells both under *in vitro* and in three-dimensional models (Kosugi et al. 2001). The most frequent ocular tumour (primary malignant) in adults is uveal melanoma. Several histone deacetylase inhibitors including romidepsin have been tested against cell lines of uveal melanoma to study their effect on apoptosis and growth of these cells. Romidepsin was found to elevate acetylation of histone H4 at specific sites. In MUM-2c, a metastatic line and two primary cell lines including M619 and OCM-1 romidepsin lowered proliferation nearly by 10%. In another cell line MUM-2b, 62% proliferation inhibition was recorded following romidepsin exposure. In total cell types excluding MUM-2c, maximum proliferation inhibition ranging from about 50% to 70% was noticed at 5 nM romidepsin. Romidepsin incited apoptosis as the number of cells in the sub-G1 fraction increased substantially compared to solvent controls. While in the negative control the sub-G1 percentage was 12.9, this percentage in cells subjected to treatment was 49.2. Activation of caspase-3 and downstream cleavage of poly (ADP-ribose) polymerase (PARP), the hallmarks of apoptosis, was confirmed from romidepsin-applied cells (Klisovic et al. 2003).

Inflammatory breast cancer (IBC) is the highest metastatic form of breast cancer that is locally advanced. This cancer type has unique characteristics, such as tumour emboli invasion into skin and fast disease advancement. Evidences suggest that romidepsin strongly destroys not only IBC tumour emboli but also the integrity of the framework of lymphatic vasculature (Klisovic et al. 2003). The impact of ras-oncogene is reversed by romidepsin use. Ras-oncogene has crucial implications in the development of tumours (Manoharan, Lehlohonolo & Magama 2019). Having the capability of transforming the ras-phenotype to a canonical one and the potential to induce cytotoxicity, the National Cancer Institute of USA premierly developed romidepsin as a compound that works against ras (Nakajima et al. 1998; Wang et al. 1998). The molecular and biological activities of romidepsin were tested on MDA-MB231 and MCF7 breast cancer cell lines. In these cells romidepsin induced apoptotic-like cell cytotoxicity, as internucleosomal degradation of DNA was confirmed from cells treated with this anticancer agent. Moreover, western immunoblotting revealed enhanced levels of phosphorylated Bcl-2 and p21Cip1 in romidepsin-incubated cells, besides activating the p33 kinase that acts on myelin basic protein (MBP). Thus, it is strongly perceptible that romidepsin-induced apoptosis is mediated through Bcl-2 phosphorylation, elevated p21CiP1 protein levels and via activation of p33 MBP kinase (Rajgolikar, Chan & Wang 1998; VanderMolen et al. 2011).

The effects of romidepsin were examined on prostate cancer cells under *in vivo* conditions. Human prostate cancer cells were implanted in severe combined immunodeficient mice that were non-obese. These mice were given romidepsin 50 mg/kg via oral route thrice weekly. Administration of romidepsin inhibited not only tumour growth but also metastasis. It was found that the mice left untreated died at 98 days while among the ones subjected to romidepsin 61% survived. Typical lung morphology was observed in treated animals. Apoptotic bodies and chromatin compaction were confirmed from the sections derived from romidepsin-treated animals. Either 50 mg/kg dose thrice a week or a single dose (1.2 g/kg) induced no pernicious effects in treated animals (Lai et al. 2008). This suggests that romidepsin has the potential to reduce prostate cancer tumour growth under *in vivo* conditions without imparting marked toxicity.

In the human HL-60 (human leukemia cell line), Caco-2 (human colon cancer) the mechanism followed by romidepsin to induce apoptosis in these cells was investigated. Following 24 hours of romidepsin intervention the cytotoxic effect accompanied with DNA fragmentation was noticed in the leukaemic cell line, while no such effect was noticed in HP100 (hydrogen peroxide (H_2O_2)-resistant sub-clone of HL-60). On the other hand, TSA, another HDAC inhibitor, induced DNA fragmentation in both HL-60 and HP100. Obstruction of HDACs by romidepsin in leukemic lines was found to a similar degree. Romidepsin treated cells manifested formation of hydrogen peroxide and alteration in the membrane potential of mitochondria. The induction of apoptosis was also observed in Caco-2 cells, which was rescued by anti-oxidant N-acetyl-cysteine (Mizutani et al. 2010). The crux from these findings is that romidepsin-induced apoptosis unlike TSA is mediated through hydrogen peroxide in leukemic lines, while in the Caco-2 line romidepsin-triggered cytotoxicity relies on enhanced reactive oxygen species generation.

Romidepsin has proved effective against solid tumours as well. Programmed Cell Death Ligand 1 (PD-L1), basically a transmembrane protein, has a critical role in a variety of malignancies. PD-L1 favours malignancies by way of reducing the immune response of the host against tumour cells (Han, Liu & Li 2020). It has been reported that HDACi have the tendency to alter the anticancer immune response states under *in vivo* conditions. Antitumour effects and the impact of romidepsin on immune responses were studied on colon cancer. Romidepsin prevented proliferation, induced cell cycle arrest and elevated apoptosis in MC38 and CT26 cells. By way of enhancing the acetylation status of histones H3 and H4 and through regulation of BRD4 (transcription factor), romidepsin increases the expression level of PD-L1. In colitis-associated cancer (CAC) mice and subcutaneous transplant tumour-bearing mice, romidepsin lowered the Th1/Th2 cell ratio. However, the FOXP3+ regulatory T cell (Tregs) showed an increase in percentage on romidepsin use (Shi et al. 2021). HDAC overexpression has been seen in biliary tract cancer and it was found that intervention of BTC cells with different HDACi alleviates their viability. Importantly BTC cells were observed to be vulnerable to romidepsin. Romidepsin lowered the activity of HDACs in these cells and induced apoptotic death. It was also evaluated that the tumours of BTC patients showed high expression of Class I isozyme HDAC2 (Mayr et al. 2021). Such patients also demonstrated brief survival. These results suggest that romidepsin may prove a promising therapeutic for treating BTC patients.

Other HDACi of the depsipeptide group are spiruchostatin A (SP-A) and spiruchostatin B (SP-B). Both these inhibitors exhibit strong HDAC inhibitory activity. The mechanism through which these inhibitors induce apoptosis in the U937 (human lymphoma) cell line was analysed. The pharmacological intervention of this cell line with the above-mentioned inhibitors induced apoptosis in this line. This was evident from enhanced DNA fragmentation, alterations in nuclear morphology. Among the two spiruchostatins, SP-B showed relatively more potential to elicit apoptosis. A substantial increase in ROS generation was proved in these U937 cells after treatment with SP-A and SP-B. The apoptotic effect induced by these inhibitors was rescued when U937 cells were subjected to N-acetyl-L-cysteine pre-treatment. Apart from this, the HDAC inhibitor treated cells showed mitochondrial membrane potential loss in addition to caspase-8 and 3 activation (Table 4.1). The further activation of Bid and an increase in cytosolic cytochrome-c that got released from the mitochondria of cancer cells was seen (Rehman et al. 2014). This discussion tempts me to speculate that SP-A and SP-B induced apoptosis is premierly

TABLE 4.1
Diverse Molecular Targets Modulated by TSA and Natural Despsipeptide Archetypes

HDAC Inhibitor	Cellular Players Inhibited	Cellular Molecules Activated
TSA	ERK1/2 phosphorylation, AKT, GSH, TRPV6 and PRMT5, cyclin D1	ROS, caspase-3, bax, p21Waf1, beclin1, p62, p-JNK and STC1, GRP94 and GRP78, caspase-9, LC3-II, p21, SOX2 and OCT4, p21Cip1,
Largazole	AKT signalling, EGFR, mTOR signalling, E2F1	Pax6 and Bdnf, p21
Romidepsin	Cyclin B1, Cdc2/Cdk, pRb, PI3K/ AKT/mTOR, demethylases, HDACs,	p21, caspase-3, p21CiP1, phosphorylated Bcl-2, ROS, PD-L1
Spiruchostatin B		ROS, caspase-8 and 3, Bid

mediated through ROS generation. These inhibitors have the ability to serve as novel therapeutics after higher degree clinical studies.

The therapeutic effects of SP-B have been ascribed to its ability to inhibit HDACs and enhance the intracellular ROS levels, especially hydrogen peroxide. The impact of a low dose of this inhibitor was studied on radiation-triggered apoptosis in the cell model of human lymphoma (U937). Combing SP-B with radiation substantially increased radiation-driven apoptosis. The increase in apoptosis correlated with enhanced levels of reactive oxygen species. The elevation in cytotoxicity due to combinatorial therapy was markedly soothed by pre-treatment of U937 cells with antioxidant, namely N-acetyl-L-cysteine. The combined regimen induced the death receptor activation in addition to the stimulation of intrinsic apoptotic pathway through facilitation of ROS-backed signalling. In other lymphoma lines such as HL-60 and Molt-4 SP-B, there was substantially escalated radiation-invoked apoptosis (Rehman et al. 2016). These results suggest that SP-B serves the function of radio-sensitizer and thus enhances radiation-provoked apoptosis through redox signalling modulation.

So, the anticancer effects of naturally derived hydroxamates, including TSA, have been taken to their logical conclusion. Moreover, light has been thrown on depsipeptides, such as largazole, spiruchostatins and largazole. The molecular signalling mechanisms elicited by these inhibitors in various preclinical models have also been taken into account. As expected, the majority of these inhibitors enhanced the acetylation degree of histones through restraining of HDACs. In certain cases, the increased production of reactive oxygen species was triggered by these inhibitors to induce cytotoxic effects in cancer cells. Cell cycle arrest, caspase activation and a decline in mitochondrial membrane potential was also noticed in other cases. In brief, hydroxamates and depsipeptides modulate multiple mechanisms, including epigenetic ones, for producing profound pharmacological effect. Plant-derived HDACi, being large in number, will be discussed in later chapters. The next segment will discuss flavonoid HDACi in a broader perspective.

REFERENCES

Al-Awadhi, F. H., Salvador-Reyes, L. A., Elsadek, L. A., Ratnayake, R., Chen, Q.-Y. & Luesch, H. (2020). Largazole is a brain-penetrant Class I HDAC inhibitor with extended applicability to glioblastoma and CNS diseases. *ACS Chemical Neuroscience 11*: 1937–1943.

Bertino, E. M. & Otterson, G. A. (2011). Romidepsin: a novel histone deacetylase inhibitor for cancer. *Expert Opinion on Investigational Drugs 20*: 1151–1158.

Campas-Moya, C. (2009). Romidepsin for the treatment of cutaneous T-cell lymphoma. *Drugs Today (Barc) 45*: 787–795.

Dai, L., He, G., Zhang, K., Guan, X., Wang, Y. & Zhang, B. (2019). Trichostatin A induces p53-dependent endoplasmic reticulum stress in human colon cancer cells. *Oncology Letters 17*: 660–667.

Emonds, E., Fitzner, B. & Jaster, R. (2010). Molecular determinants of the antitumor effects of trichostatin A in pancreatic cancer cells. *World J Gastroenterol 16*: 1970–1978.

Fennell, K. A., Möllmann, U. & Miller, M. J. (2008). Syntheses and biological activity of amamistatin B and analogs. *J Org Chem 73*: 1018–1024.

Gilardini Montani, M. S., Granato, M., Santoni, C., Del Porto, P., Merendino, N., D'Orazi, G., Faggioni, A. & Cirone, M. (2017). Histone deacetylase inhibitors VPA and TSA induce apoptosis and autophagy in pancreatic cancer cells. *Cell Oncol (Dordr) 40*: 167–180.

Han, Y., Liu, D. & Li, L. (2020). PD-1/PD-L1 pathway: current researches in cancer. *American Journal of Cancer Research 10*: 727–742.

Jeong, K., Kim, S. J., Oh, Y., Kim, H., Lee, Y. S., Kwon, B. S., Park, S., Park, K. C., Yoon, K. S., Kim, S. S., Ha, J., Kang, I. & Choe, W. (2014). p53 negatively regulates Pin1 expression under ER stress. *Biochem Biophys Res Commun 454*: 518–523.

Klisovic, D. D., Katz, S. E., Effron, D., Klisovic, M. I., Wickham, J., Parthun, M. R., Guimond, M. & Marcucci, G. (2003). Depsipeptide (FR901228) inhibits proliferation and induces apoptosis in primary and metastatic human uveal melanoma cell lines. *Investigative Ophthalmology & Visual Science 44*: 2390–2398.

Kosugi, H., Ito, M., Yamamoto, Y., Towatari, M., Ito, M., Ueda, R., Saito, H. & Naoe, T. (2001). In vivo effects of a histone deacetylase inhibitor, FK228, on human acute promyelocytic leukemia in NOD / Shi-scid/scid mice. *Jpn J Cancer Res 92*: 529–536.

Lai, M. T., Yang, C. C., Lin, T. Y., Tsai, F. J. & Chen, W. C. (2008). Depsipeptide (FK228) inhibits growth of human prostate cancer cells. *Urol Oncol 26*: 182–189.

Lambert, I. H., Nielsen, D. & Stürup, S. (2020). Impact of the histone deacetylase inhibitor trichostatin A on active uptake, volume-sensitive release of taurine, and cell fate in human ovarian cancer cells. *American Journal of Physiology-Cell Physiology 318*: C581–C597.

Liu, J. H., Cao, Y. M., Rong, Z. P., Ding, J. & Pan, X. (2021). Trichostatin A induces autophagy in cervical cancer cells by regulating the PRMT5-STC1-TRPV6-JNK pathway. *Pharmacology 106*: 60–69.

Liu, Y., Salvador, L. A., Byeon, S., Ying, Y., Kwan, J. C., Law, B. K., Hong, J. & Luesch, H. (2010). Anticolon cancer activity of largazole, a marine-derived tunable histone deacetylase inhibitor. *The Journal of Pharmacology and Experimental Therapeutics 335*: 351–361.

Manoharan, K., Lehlohonolo, I. & Magama, S. (2019). DPPH radical scavenging activity of extracts from Rhamnus prinoides. *Journal of Medicinal Plants Research 13*: 329–334.

Mayr, C., Kiesslich, T., Erber, S., Bekric, D., Dobias, H., Beyreis, M., Ritter, M., Jäger, T., Neumayer, B., Winkelmann, P., Klieser, E. & Neureiter, D. (2021). HDAC screening identifies the HDAC Class I inhibitor romidepsin as a promising epigenetic drug for biliary tract cancer. *Cancers (Basel) 13*: 3862.

Mizutani, H., Hiraku, Y., Tada-Oikawa, S., Murata, M., Ikemura, K., Iwamoto, T., Kagawa, Y., Okuda, M. & Kawanishi, S. (2010). Romidepsin (FK228), a potent histone deacetylase inhibitor, induces apoptosis through the generation of hydrogen peroxide. *Cancer Sci 101*: 2214–2219.

Nakajima, H., Kim, Y. B., Terano, H., Yoshida, M. & Horinouchi, S. (1998). FR901228, a potent antitumor antibiotic, is a novel histone deacetylase inhibitor. *Exp Cell Res 241*: 126–133.

Prince, H. M. & Dickinson, M. (2012). Romidepsin for cutaneous T-cell lymphoma. *Clin Cancer Res 18*: 3509–3515.

Rajgolikar, G., Chan, K. K. & Wang, H. C. (1998). Effects of a novel antitumor depsipeptide, FR901228, on human breast cancer cells. *Breast Cancer Res Treat 51*: 29–38.

Rehman, M. U., Jawaid, P., Yoshihisa, Y., Li, P., Zhao, Q. L., Narita, K., Katoh, T., Kondo, T. & Shimizu, T. (2014). Spiruchostatin A and B, novel histone deacetylase inhibitors, induce apoptosis through reactive oxygen species-mitochondria pathway in human lymphoma U937 cells. *Chem Biol Interact 221*: 24–34.

Rehman, M. U., Jawaid, P., Zhao, Q., Li, P., Narita, K., Katoh, T., Shimizu, T. & Kondo, T. (2016). Low-dose spiruchostatin-B, a potent histone deacetylase inhibitor enhances radiation-induced apoptosis in human lymphoma U937 cells via modulation of redox signaling. *Free Radical Research 50*: 1–15.

Reid, B. M., Permuth, J. B. & Sellers, T. A. (2017). Epidemiology of ovarian cancer: a review. *Cancer Biol Med 14*: 9–32.

Rhodes, L. V., Nitschke, A. M., Segar, H. C., Martin, E. C., Driver, J. L., Elliott, S., Nam, S. Y., Li, M., Nephew, K. P., Burow, M. E. & Collins-Burow, B. M. (2012). The histone deacetylase inhibitor trichostatin A alters microRNA expression profiles in apoptosis-resistant breast cancer cells. *Oncol Rep 27*: 10–16.

Shao, G., Lai, W., Wan, X., Xue, J., Wei, Y., Jin, J., Zhang, L., Lin, Q., Shao, Q. & Zou, S. (2017). Inactivation of EGFR/AKT signaling enhances TSA-induced ovarian cancer cell differentiation. *Oncol Rep 37*: 2891–2896.

Shi, Y., Fu, Y., Zhang, X., Zhao, G., Yao, Y., Guo, Y., Ma, G., Bai, S. & Li, H. (2021). Romidepsin (FK228) regulates the expression of the immune checkpoint ligand PD-L1 and suppresses cellular immune functions in colon cancer. *Cancer Immunology, Immunotherapy 70*: 61–73.

Sun, S., Han, Y., Liu, J., Fang, Y., Tian, Y., Zhou, J., Ma, D. & Wu, P. (2014). Trichostatin A targets the mitochondrial respiratory chain, increasing mitochondrial reactive oxygen species production to trigger apoptosis in human breast cancer cells. *PLoS One 9*: e91610.

Valdez, B. C., Brammer, J. E., Li, Y., Murray, D., Liu, Y., Hosing, C., Nieto, Y., Champlin, R. E. & Andersson, B. S. (2015). Romidepsin targets multiple survival signaling pathways in malignant T cells. *Blood Cancer Journal 5*: e357-e357.

VanderMolen, K. M., McCulloch, W., Pearce, C. J. & Oberlies, N. H. (2011). Romidepsin (Istodax, NSC 630176, FR901228, FK228, depsipeptide): a natural product recently approved for cutaneous T-cell lymphoma. *J Antibiot (Tokyo) 64*: 525–531.

Vigushin, D. M., Ali, S., Pace, P. E., Mirsaidi, N., Ito, K., Adcock, I. & Coombes, R. C. (2001). Trichostatin A is a histone deacetylase inhibitor with potent antitumor activity against breast cancer in vivo. *Clin Cancer Res 7*: 971–976.

Vinodhkumar, R., Song, Y. S. & Devaki, T. (2008). Romidepsin (depsipeptide) induced cell cycle arrest, apoptosis and histone hyperacetylation in lung carcinoma cells (A549) are associated with increase in p21 and hypophosphorylated retinoblastoma proteins expression. *Biomed Pharmacother 62*: 85–93.

Wang, M., Sun, X. Y., Zhou, Y. C., Zhang, K. J., Lu, Y. Z., Liu, J., Huang, Y. C., Wang, G. Z., Jiang, S. & Zhou, G. B. (2020). Suppression of Musashi-2 by the small compound largazole exerts inhibitory effects on malignant cells. *Int J Oncol 56*: 1274–1283.

Wang, R., Brunner, T., Zhang, L. & Shi, Y. (1998). Fungal metabolite FR901228 inhibits c-Myc and Fas ligand expression. *Oncogene 17*: 1503–1508.

Wu, L.-C., Wen, Z.-S., Qiu, Y.-T., Chen, X.-Q., Chen, H.-B., Wei, M.-M., Liu, Z., Jiang, S. & Zhou, G.-B. (2013). Largazole arrests cell cycle at G1 phase and triggers proteasomal degradation of E2F1 in lung cancer cells. *ACS Med Chem Lett 4*: 921–926.

You, B. R. & Park, W. H. (2013). Trichostatin A induces apoptotic cell death of HeLa cells in a Bcl-2 and oxidative stress-dependent manner. *Int J Oncol 42*: 359–366.

5 Promising Therapeutic Benefits of Flavonoid Histone Deacetylase Inhibitors with Special Emphasis on Modulation of Central Molecular Mechanisms

Inhibitors of histone deacetylases derived from plant sources are largely emerging as encouraging molecules for targeting different cancer types. They have the ability to serve as potential substitutes to synthetic histone deacetylase inhibitors (HDACi). Based on chemotype, these inhibitors may be flavonoids, isothiocyanates, organosulfur compounds, short chain fatty acids, stilbenes, polyketides, bromotyrosine and coumarin derivatives. The majority of plant derived HDACi are currently under preclinical studies, some are undergoing clinical trials while none of the plant derived HDACi is presently FDA approved. This chapter will focus wholly on HDACi from plant sources belonging to the flavonoid group, their classification and their optimistic anticancer activities against diverse cancer models under in vitro and in vivo environment. Further, the outcome of clinical studies involving these inhibitors will also be taken into consideration.

5.1 SHORT INTRODUCTION TO PLANT-DERIVED HDAC INHIBITORS

Substantial toxicities accompanied with the therapeutic intervention of synthetic HDACi have motivated the scientific community to concentrate on plant-derived molecules possessing HDAC obstructing ability (Evans & Ferguson 2018). Due to the intense efforts of different research labs globally, substantial numbers of molecules identified from a variety of plant sources have been certified to have HDAC inhibitory potential (Son et al. 2007b; Pal-Bhadra et al. 2012). Among the plant-extracted HDACi, the majority come under the umbrella of flavonoids (Ganai et al. 2021). The rest of the inhibitors are distributed among other groups, such as prenylated isoflavones, organosulfur compounds, isothiocyanates, short chain fatty acids, hydroxycinnamic acids, hydroxybenzoic acids, quinones, stilbenes, alkaloids and curcuminoids (Son et al. 2007b; Waldecker et al. 2008; Brown, Allgar & Wong 2016). As the flavonoid group of HDACi is vast, it will be interesting to discuss different subgroups of flavonoids and their corresponding molecules.

5.2 CLASSIFICATION OF PHENOLIC COMPOUNDS AND FLAVONOIDS

It has been estimated that more than 5,000 distinct flavonoids have been characterized from plants (Havsteen 2002). They have been reported from the plant kingdom in general and from plant cells associated with photosynthesis in particular. Phenolic compounds may be simple phenols or polyphenols. Tannins and flavonoids come under the banner of polyphenols (Figure 5.1) (Robbins & Bean 2004; Long et al. 2014). Flavonoids contain a flavan nucleus made of 15 carbon atoms organized in three rings (diphenylpropane skeleton). These rings are named A, B and C rings. In general, flavonoids exist in plant in the form of glycosylated derivatives and contribute significantly to the alluring colours of fruits, leaves and flowers. Based on the C-ring carbon to which the B-ring is linked and the level of C-ring unsaturation and oxidation, flavonoids have been divided into several groups (Pal & Saha 2013). These groups encompass isoflavones, flavanones, flavonols, flavones, flavan-3-ols

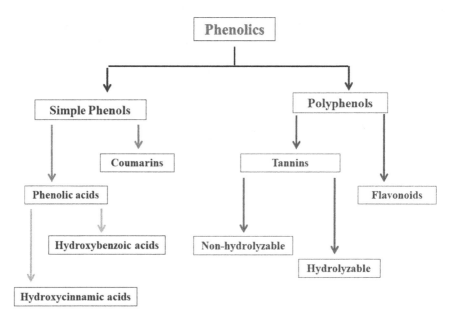

FIGURE 5.1 An easy-to-understand classification of phenolic compounds from natural sources. Phenols may be either simple phenols or polyphenols. Simple phenols cover coumarins and phenolic acids. These phenolic acids are further subdivided into hydroxycinnamic acids and hydroxybenzoic acids. Polyphenols fence in tannins and flavonoids. Some tannins are hydrolyzable while others are non-hydrolyzable.

and anthocyanins. These groups will be discussed individually and the HDACi falling within the confines of these groups will also be discussed.

5.3 EXTENSIVE COMPENDIUM OF FLAVONOID HDAC INHIBITORS

Isoflavones are the flavonoids in which the B-ring is attached to the C-ring at position 3. Genistein, daidzein and glycitein are prominent isoflavones. Certain isoflavones including daidzein and genistein are deemed phytoestrogens as they have demonstrated estrogenic activity in *in vivo* models (Szkudelska & Nogowski 2007; Panche, Diwan & Chandra 2016; Desmawati & Sulastri 2019). Flavanones also termed dihydroflavones possess the saturated C-ring. They differ from flavones in lacking the double bonding in between position 2 and 3 of defined ring (Khan, Zill & Dangles 2014). Naringenin, naringin, hesperidin, sylibin and eriodictyol are prominent examples of flavanones. On the other hand, flavonoids possessing the ketone function (group) are referred to as flavonols. Flavonols serve as the building blocks for proanthocyanins. Quercetin, kaempferol, isorhamnetin, fisetin and myricetin are among the extensively studied flavonols (Desmawati & Sulastri 2019). On the C-ring at position 3, flavonols contain the hydroxyl group in comparison to flavones (Graf, Milbury & Blumberg 2005). Flavones form the subgroup of flavonoids and have manifested substantial biological activity under both *in vitro* and *in vivo* conditions (Hostetler, Ralston & Schwartz 2017). They vary from other flavonoid subgroups in terms of having an unsaturation (double bond) in the flavonoid skeleton. This double bond lies between C2 and C3 of this skeleton and no substitution occurs at C3 position. Apart from this, at C4 location flavones are oxidized (Martens & Mithöfer 2005). Apigenin, luteolin, chrysin, rutin and wogonin are some of the famous examples of flavones that have shown significant results in various preclinical studies (Chi et al. 2003; Wojdyło, Oszmiański & Czemerys 2007; Gullón et al. 2017; Imran et al. 2019; Garg & Chaturvedi 2022).

Flavanols are regarded as the 3-hydroxy derivatives of so-called flavanones. In other words, they are also called dihydroflavonols. As in flavanols, the hydroxyl group is inevitably linked to location 3 of the C-ring, and they are also called as flavan-3-ols. No double bond exists between position 2 and the successive position (3), unlike many flavonoids. Epicatechin, catechin, epigallocatechin gallate, epigallocatechin are the few examples of flavanols (Pal & Saha 2013; Farooqui & Farooqui 2018). Further, Chalcones are considered as another subgroup of the flavonoid family. These secondary metabolites are ubiquitous in medicinal and edible plants. Chalcones share a usual scaffold (1, 3-diaryl-2-propen-1-one), which is also termed as chalconoid. This scaffold exhibits both trans and cis forms, but the former is favoured due to its high thermodynamic stability (Sahu et al. 2012; Rudrapal et al. 2021). Chemically, parent chalcones possess α, β-unsaturated carbonyl (3-carbon)-system that joins the two aromatic rings. Chalcones have shown distinct properties including anticancer, antioxidant, antiviral, antiinflammatory and other activities (Rudrapal et al. 2021). They serve as the biogenetic antecedents of flavonoids. Certain chalcones characterized from plant sources encompass butein, flavokawin, 4-hydroxyderricin, cardamonin and many more (Valavanidis & Vlachogianni 2013). Another class of flavonoids is

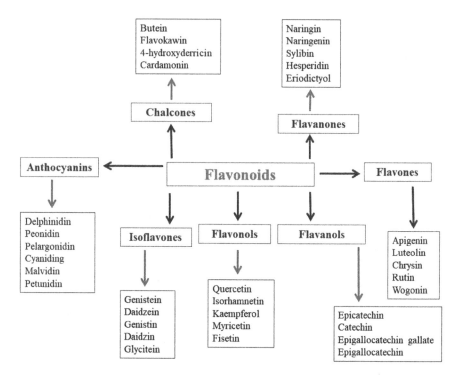

FIGURE 5.2 Overview of different types of flavonoids and their corresponding examples. Flavonoids may be flavones, flavanols, flavonols, flavanones, chalcones, isoflavones or anthocyanins. While apigenin, chrysin and luteolin come under flavones; epicatechin falls within the flavanols boundary; quercetin under flavonols; delphinidin is the representative of anthocyanins; naringin and naringenin are the typical flavanones while butein represents chalcones.

constituted by anthocyanins, which are water-soluble and natural pigments that exist in plants as glycosides. They are regarded as polymethoxy or polyhydroxy offshoots of 2-phenylbenzophyryllium (Calogero et al. 2013). Among the 17 anthocyanins, only six are extensively distributed. These six broadly distributed anthocyanins include delphinidin, cyaniding, peonidin, petunidin, malvidin and pelargonidin (Patel, Jain & Patel 2013). These six anthocyanins vary from each other at position 5´ and 3´ of the ring B (Khoo et al. 2017). As the family of phenolic compounds is broad, it would be interesting to put their overview simply in schematic form (Figure 5.2).

5.4 ANTINEOPLASTIC EFFECT OF ISOFLAVONOID HDAC INHIBITORS

Vegetables and fruits, due to their polyphenol content, alleviate the risk of cancer. Polyphenols present in various functional foods target the epigenetic regulatory enzymes and thus repeal the aberrant epigenetic signatures. The effect of isoflavone has been inspected using human cervical cancer cells as models. Genistein lowered the expression and activity of HDACs and DNA methyltransferases (DNMTs) in cervical cancer (HeLa) cells. It has been predicted through molecular modelling that genistein can interact with a variety of members of HDAC families and DNMTs. Genistein de-repressed epigenetically silenced tumour suppressor genes by reversing the methylation at their promoter sites (Sundaram et al. 2018). The impact of genistein was examined on different molecular targets of HeLa cells. Genistein was found to modulate the genes involved in different cellular processes such as migration, cell cycle regulation and inflammation. Apart from these, MAPK and PI3K pathways were modulated by genistein. Among the modulated genes *MMP14, TWIST1, TERT, CCNB1* and others were conspicuous. Expression of HDACs, DNMTs, histone methyltransferases (HMTs) and histone phosphorylases were also modulated by genistein intervention. Further, the activity of the first three enzymes mentioned above was reduced by genistein. Reduction in DNA methylation at the global level was noticed following genistein treatment. Methylation at the promoter regions of various tumour suppressor genes, namely *PTEN, RUNX3, FHIT and SOC5*, was lowered by this isoflavone (Sundaram et al. 2019).

Epigenetic modifications such as histone modifications and DNA methylation are critical for the normal functioning of cells. The induction of cancer occurs due to deacetylation of the promoter regions of tumour suppressor genes and subsequent hypermethylation. DNMT1, DNMT3a and last, but no way least, DNMT3b belong to a family of DNMTs and cause DNA hyhpermethylation. Genistein, the DNA demethylating agent, has shown anticancer activity against hepatocellular carcinoma cell models. It was found to trigger programmed cell death and thwart the proliferation of hepatocellular carcinoma cells. Genistein obstructed cell growth, provoked apoptosis and induced the expression of the above-mentioned members of the DNMT family. The effect of genistein on gene expression was recorded to be more substantial than the well-known HDAC inhibitor TSA (Sanaei et al. 2018).

Drug resistance in cholangiocarcinoma occurs in certain patients due to which they show relapse. Daidzein, an isoflavone, has gathered the focus of the scientific community as it has demonstrated antiproliferative effect against cancer cells. This isoflavone inhibited cell growth in cholangiocarcinoma cells and induced G1 arrest. Daidzein lowered the expression of several molecular players including PCNA, c-myc and cyclin D1 but elevated the p21 expression. Phosphorylated ERK expression was alleviated due to which its shuttling to nucleus was impeded. The daidzein-induced effects on proliferation of cholangiocarcinoma cells were rescued on using ceramide C6, an ERK agonist. Further, the antiproliferative activity of daidzein was also seen in, *in vivo* xenografts. This was accompanied with p-ERK, PCNA and c-myc down-regulation (Zheng et al. 2017). Thus, daidzein-invoked antiproliferative effect may be attributed to inhibition of ERK pathway and arrest of cell cycle at the above-mentioned phase.

Daidzein induced apoptosis markedly in ovarian cancer cells and bridled their migration, invasion in addition to proliferation. Besides, daidzein lowered cyclin D1 expression, p21, p-GSK3β, p-AKT, p-FAK and PI3K. A similar effect was shown by genistein on ovarian cancer cells (Chan et al. 2018). From this, it is clear that daidzein and genistein modulates multiple players in ovarian cancer cells for bringing therapeutic effect. Ovarian cancer is the premier cause of deaths due to gynaecological cancers. At first this cancer responds to chemotherapy, but afterwards quite often relapse occurs. Daidzein has manifested multiple pharmacological characteristics including anticancer and antimicrobial activity. The effect of this isoflavone type inhibitor was explored against one normal ovarian cell line and a series of human ovarian cancer cells. Daidzein demonstrated strong anticancer potential against the SKOV3 model at micromolar concentration. On the other hand, daidzein demonstrated relatively lower effect on Moody cells, the normal ovarian cells. In SKOV3 cells, pharmacological intervention with daidzein changed the morphology of these cells and induced mitochondria-mediated apoptosis. Daidzein-instigated apoptosis was mediated by mitochondria and was caspase-dependent. Certain molecular players, including cyclin B1, pCdc2, Cdc2, Cdc25c and pCdc25c, were found to undergo decline in expression after daidzein exposure. The inhibition of cell migration was also confirmed from treated cells and this inhibition was possibly due to the mitigated expression profile of matrix metalloproteinases such as (MMP)-2 and -9. Cell growth inhibition due to daidzein has been ascribed to its ability to hamper the expression level of p-MAPKK and p-SRK. Moreover, daidzein showed a significant reduction in tumorigenesis in mouse xenograft models bearing the SKOV3 cell tumour (Hua et al. 2018). Together these findings suggest daidzein has an antiovarian cancer property that needs to be explored further on clinical models.

Daidzein has been tested against human breast cancer cell line MCF-7 and it was found that these cells are sensitive to this isoflavone. The induction of apoptosis was noticed in these cells and a 1.4 fold increase in caspase 3/7 was seen in daidzein treated cells. Additionally, in treated cells down-regulation of *Bcl2* and enhanced expression of Bax was evinced. Apoptosis in daidzein-added cells was triggered due to increased production of ROS, signifying the intrinsic apoptotic pathway as the cause of cell cytotoxicity. While the elevated expression of ERβ reduced ERα expression was revealed in cells subjected to intervention (Kumar & Chauhan 2021). The gist of these results is that daidzein induces apoptotic signally in breast cancer cells through increased generation of ROS and by reducing the ration of ERα expression/ ERβ. Another study scrutinized the impact of daidzein on MCF-7 cells. It was revealed that daidzein exerts antiproliferative property in time and concentration dependent fashion. Production of ROS due to daidzein was the critical initiator factor for apoptotic signalling. ROS escalation was associated with mitochondrial membrane potential alteration, bcl-2 down-regulation and bax increase. This resulted in the release of cytochrome C into the cytosol from mitochondria. Following this, activation of two caspases, namely caspase-9 and 7, caused eventual cell death. ROS-mediated and caspase-dependent cell death on daidzein use was certified through use of N-acetyl-L-cysteine, a known antioxidant and z-VAD-fmk, a pan-caspase inhibitor (Jin et al. 2010). In human colon cancer (HT-29) cells it has been reported that genistein induces apoptosis more effectively than daidzein. Both these isoflavones lowered the accumulation of lipid droplets and pacified the expression of ADRP, Perilipin-1, Tip-47 family proteins and levels of vimentin. Besides, these plant-derived modulators increased the expression of Fas, glycerol-3-phosphate acyltransferase (GPAT3), PPAR-γ and FABP. Genistein and daidzein reduced the expression of PI3K while enhancing the caspase-8 and FOXO3a markedly (Liang et al. 2018).

TABLE 5.1
Summary of Molecular Targets Modulated by Isoflavonoids for Inducing Apoptosis in Different Cancer Types

Isoflavonoid	Cancer Type	Cellular Targets Inhibited	Targets Activated
Genistein	Cervical	HDACs, DNMTs	PTEN, RUNX3, FHIT and SOC5
	Hepatocellular		DNMT1, DNMT3a, DNMT3b
	Colon	Tip-47, ADRP, Perilipin-1	Fas, FABP, PPAR-γ
Daidzein	Cholangiocarcinoma	c-myc, cyclin D1, PCNA, p-ERK	p21
	Ovarian	cyclin D1, GSK3β, PI3K, p-AKT Cdc2, Cdc25c, cyclin B1, pCdc2, pCdc25c, MMP-2, MMP-9, p-SRK, p-MAPKK	
	Breast	Bcl2, ER α , ROS	Caspase 3/7, Bax, ER β
		ROS, bcl-2	Bax, caspase-9, caspase-7
Glycitein	Gastric	ROS, NF-κB	MAPK, STAT3

Glycitein constitutes about 5–10% of the entire isoflavones in soy products. While investigating this isoflavone on human breast carcinoma cell line (SKBR-3), it was found that low concentrations of this flavone induce cytostatic effect and higher concentrations were cytotoxic. Glycitein was demonstrated to damage the plasma membrane of these cells and subsequent enhancement of membrane permeability (Zhang et al. 2015). Mounting evidence endorses the antiproliferative action of glycitein against human prostate cancer and human breast cancer cells. Only recently, the underlying molecular mechanism involved in glycitein-induced anticancer effect has been unravelled. This plant isoflavone manifested substantial cytotoxicity against human gastric adenocarcinoma AGS cells. The reduction in transmembrane mitochondrial potential and induction of cell cycle arrest was evident from cells treated with this soy-derived inhibitor. Glycitein favoured ROS production resulting in the activation of mitogen-activated protein kinase (MAPK) and blockade of STAT3 and NF-κB signalling cascades. N-acetyl-L-cysteine treatment or the use of MAPK inhibitor restored the STAT3 and NF-κB (Zang et al. 2019). Speaking in few words, AGS cell apoptosis induced by glycitein is mediated by ROS by way of activating MAPK and consequent inactivation of STAT3 and NF-κB signalling mechanisms (Table 5.1).

5.5 FLAVANONES IN OBSTRUCTING CANCER PROGRESSION

Like isoflavones, flavanones have also shown strong anticancer activity against pre-clinical models of distinct cancers. For instance, naringenin exhibits antitumour and antiinflammatory, antioxidant and antidiabetic properties (Rani et al. 2016; Zaidun, Thent & Latiff 2018; Den Hartogh & Tsiani 2019). The migration of a variety of cancerous cell lines is hampered by this flavanone (Zhang et al. 2016; Chang et al. 2017; Wang et al. 2019). Its effects have been explored in lung cancer cell lines, namely A549. A noticeable change in proliferation of these cells was seen in naringenin post-treatment. Naringenin restrained migration of A549 cells in a dose-dependent manner. Through zymography it was delineated that this inhibitor blocked MMP-2 and MM-9 function. Inhibition of AKT function dose-dependently was another effect observed with treatment of naringenin (Chang et al. 2017). These results tempt me to speculate that the naringenin-induced pharmacological effect on lung cancer cells is mediated by way of obstruction of AKT, MMP-2 and MMP-9 activities.

Bladder cancer is considered to have a high death rate because of its metastatic potential. The molecular mechanism modulated by naringenin for inhibiting migration of bladder cancer cells was studied using TSGH-8301 cells as models. The reduction in cell viability was quantified while using naringenin on these cells in micromolar concentrations for 24 hours duration. Zymography coupled to western blot revealed lowered expression of MMP-2 in treated TSGH-8301 cells. The exposure of these cells to naringenin was accompanied with mitigated activity of AKT (Liao et al. 2014). Thus, the ability of naringenin to block metastasis of bladder cancer cells is strongly backed by its tendency to hamper MMP-2 and AKT activities.

Another flavanone, naringin, has proved effective against multiple cancers. It impedes growth of cancer cells, induces cell cycle arrest, modulates oxidative stress, angiogenesis and inflammation. The viability and proliferation of transitional cell carcinoma cells was alleviated by naringin under *in vitro* conditions (Oršolić, Štajcar & Bašić 2009). Triple-negative breast cancer is a serious clinical concern due to its bellicose behaviour, mediocre prognosis and absence of targeted therapies. Naringin has been evaluated against triple-negative breast cancer cells and *in vivo* models. This molecule showed multiple effects on these cell lines such as inhibited their proliferation, facilitated apoptosis, lowered survivin and enhanced p21. Further, naringin intervention was associated with inhibition of the β-catenin signaling pathway as overexpression of β-catenin reinstated the naringin-induced effects. In MDA-MB-231 xenograft mice, naringin exerted an antitumour effect that was mediated through diminished expression of active β-catenin and survivin (Li et al. 2013). In short, naringin inhibits the β-catenin pathway for eliciting cytotoxic effect in triple-negative breast cancer models.

The pharmacological effects of naringin were studied on human cervical cancer (HeLa) cells. Naringin incorporation reduced cell viability resulting in inhibition of growth. Additionally, induction of apoptosis and cleavage of caspase-3 was noticeable in treated cells. Naringin addition weakened the expression profile of cyclooxygenase-2 (COX-2), phosphorylated (p) nuclear factor κB (NF-κB) p65 subunit and cysteinyl aspartate proteinase-1 (caspase-1). Therapeutic intervention involving NF-κB inhibitor (PTDC), or COX-2 inhibitor (NS-398) or caspase-1 inhibitor (SC-3069) reproduced the effect of naringenin thereby confirming that the effect of the latter is mediated via these molecular players (Zeng et al. 2014). Speaking concisely, naringin-incited apoptotic effect in cervical cancer cells follows the NF-κB/COX-2-caspase-1 pathway.

TABLE 5.2
Molecular Players Influenced by Flavanones for Provoking Cytotoxic Effect in Diverse Cancer Models

Flavanones	Name of Cancer	Targets Hampered	Molecules Promoted
Naringenin	Lung cancer	MMP-9, MMP-2, AKT	
	Bladder cancer	MMP-2, AKT	
Naringin	Triple-negative breast cancer	Survivin, β-catenin	p21
	Cervical cancer	COX-2, p-NF-κB	
Hesperidin	Hepatic cancer		ERK1/2, Caspase
	Gastric cancer	BCL2	Bax, CASP3

The therapeutic benefits of hesperidin, another member of the flavanone class, have been unravelled using human hepatic cancer (HepG2) cells as models. Hesperidin was found to be comparatively better than neohesperidin and naringin in inducing cytotoxic effect in these malignant cells (Banjerdpongchai et al. 2016). Hesperidin-provoked apoptosis in HepG2 cells followed the mitochondrial as well as death receptor pathway.

Hesperidin occurring in *Citrus* is typical for its antiinflammatory, anticancer and antioxidant properties. Hesperidin evoked a specific type of programmed cell death in HepG2 cells. Treatment with hesperidin resulted in swelling of mitochondria and endoplasmic reticulum, cytoplasmic vacuolization and uncondensed chromatin formation. Hesperidin-induced death was not accompanied with DNA fragmentation, caspase activation and subsequent cleavage of PARP. This death was like paraptosis and was interceded through ERK1/2 activation (Yumnam et al. 2014).

In a variety of cancer lines, hesperidin has been proved to lower the proliferation. The effects of hesperidin have been assessed on human gastric cancer (SNU-668) cells. Hesperidin caused apoptotic death in these cells that was escorted with downregulation of B-cell CLL/lymphoma 2 (BCL2) and up-regulation of BCL2-associated X protein (BAX). Further, in treated cells not only expression but also the activity of the main apoptotic factor caspase3 (CASP3) was enhanced (Park et al. 2007). This outcome suggests that hesperidin-goaded apoptosis occurs via caspase-3 activation route (Table 5.2).

5.6 ANTINEOPLASTIC PROPERTY OF FLAVONOLS

Flavonols such as quercetin, kaempferol and others have manifested anticancer activity in multiple cancers. It will be interesting to discuss the anticancer potential of these flavonols individually.

5.6.1 ANTICANCER POTENTIAL OF QUERCETIN

Quercetin has offered positive benefits for preventing cancer. The epigenetic mechanism followed by this flavone for induction of apoptosis has been dissected. In human leukemia cells, quercetin aroused extrinsic apoptotic signalling via caspase-8 activation and cleavage of Bid and cytochrome-C release. This inhibitor facilitated activation of c-Jun, enhanced core histone H3 acetylation, which in turn resulted in FasL induction. Quercetin-promoted H3 acetylation was attributed to its ability to activate histone acetyltransferases (HATs) and its inhibitory effect on HDACs (Lee, Chen & Tseng 2011). This proves that quercetin-kindled apoptosis in HL-60 is partly mediated through epigenetic mechanism. Breast cancer stem cell self-renewal, proliferation and invasiveness were found to be inhibited by quercetin. The degree of expression of various proteins favouring cancer progression and tumorigenesis including dehydrogenase 1A1, mucin 1 and C-X-C chemokine receptor type 4 was attenuated by this flavonol (Wang et al. 2018).

Studies related to quercetin and its influence on proliferation of two breast cancer cell lines, namely MDA-MB-231 and MCF-7, has been experimented. While substantial cytotoxic effect was seen in the latter cell line, no such effect was recorded in the former cell line. Quercetin exposure in MCF-7 cells resulted in G1 arrest, apoptosis, down-regulation of p21, cyclin D1 and phospho-p38MAPK and Twist (Ranganathan, Halagowder & Sivasithambaram 2015). Conclusively, quercetin-aroused apoptotic death in breast cancer cells is mediated through Twist inhibition by way of p38MAPK.

The anticancer strength of flavonol quercetin was estimated on nine different cancer cell lines ranging from colon carcinoma to ovarian cancer. This inhibitor was able to induce apoptotic signalling in all the concerned cancer lines. In comparison to the control, marked apoptosis induction was visible from Raji cell lines, LNCaP, MOLT-4 and CT-26. CT-26 and MCF-7 tumour bearing mice demonstrated statistically substantial shrinkage in tumour dimensions (Hashemzaei et al. 2017). This establishes that quercetin has the potential to kill cancer cells not only under *in vitro* conditions but also under *in vivo* conditions. Quercetin by way of inducing apoptosis in pancreatic cancer cells hampered their growth (Angst et al. 2013). Tumour growth reduction was also noticed in the pancreatic cancer xenograft model generated orthotopically. Pancreatic ductal adenocarcinoma has high mortality and poor prognosis. For its progression and maintenance unusual hedgehog signalling has a critical role. Quercetin offers therapeutic benefit in pancreatic ductal adenocarcinoma by tuning sonic hedgehog signalling. Proliferation of pancreatic cancer cells was blocked by quercetin. This effect was mediated through c-Myc down-regulation. Quercetin lowered TGF-β1 and thus restrained migration and invasion of these cells. Overexpression of sonic hedgehog protein partially reversed the quercetin-induced effects in pancreatic cancer cells. Atypical expression of sonic hedgehog triggers activation of TGF-β1/Smad2/3 signaling facilitating epithelial to mesenchymal transition through induction of snail1 and Zeb2 (Guo et al. 2021). These results tempt me to conclude that quercetin impedes migration, invasion, growth and induces apoptotic death in pancreatic cancer cells by negatively regulating sonic hedgehog.

The therapeutic effects of quercetin have been studied in the context of oral squamous carcinoma involving SCC-9 as models. Time and dose-dependent suppression of cell growth was seen on quercetin intervention. The membrane integrity and morphology of these cells changed after treatment with this flavonol. While treatment up to 48 hours induced necrosis, longer exposure (72 hours) induced apoptosis that was evident from activation of caspase-3. Cell cycle arrest was observed in S phase following quercetin exposure and was associated with inhibition of thymidylate synthase, the critical enzyme involved in S phase (Haghiac & Walle 2005). Thus, the short duration exposure of SCC-9 cells to quercetin induces necrosis while longer exposure culminates in apoptosis. Apoptosis in these cells occurs through reduction in the protein levels of thymidylate synthase. The mind-blowing study on oral cancer (SAS) cells again revealed the anticancer property of quercetin. Quercetin-elicited death in these cells in time-dependent fashion enhanced ROS generation and production of Ca^{2+}, reduced transmembrane potential of mitochondria, enhanced the number of apoptotic cells besides modulating the degree of expression of apoptosis-associated protein in these cells. Cells collected from quercetin-treated plates were estimated to have elevated protein levels of Fas, caspase-8 and Fas-associated protein death domain (FADD), quercetin-induced endoplasmic reticulum stress in SAS cells as the degree of expression of gastrin-releasing peptide-78, activating transcription factor (ATF)-6α and ATF-6β (hallmarks of ER stress) was elevated. The expression of certain molecular players related to apoptosis was escalated. These proteins encompass endonuclease G, apoptosis-inducing factor and cytochrome c (Ma et al. 2018; Mirazimi et al. 2022). Thus, the death of oral cancer cells due to quercetin occurs through multiple mechanisms and quercetin may prove as a promising flavonol for circumventing this cancer.

In tumour cells, enhanced glycolysis elevates tumour progression risk and subsequent mortality. Thus, the inhibition of cancer growth and advancement may be achieved through the blockade of glycolysis. Hexokinase-2 has great importance in glucose metabolism as its enhanced expression facilitates the tumour cell growth, which makes it a critical molecular target for anticancer drugs and therapy. Quercetin exhibited intense anticancer effect against hepatocellular carcinoma cells by way of altering aerobic glycolysis. In these cells quercetin lowered the hexokinase-2 protein levels and obstructed AKT/mTOR pathway. Further, this inhibition of hexokinase-2 expression was also observed in hepatocellular carcinoma xenografts (Wu et al. 2019). This means patients' hexokinase-2 driven hepatocellular carcinoma can be subdued by intervention involving quercetin.

Quercetin was tested against two human acute myeloid leukemia xenograft models and on two cell lines U937 and HL-60. While probing the underlying molecular mechanism modulated by quercetin in these lines and xenograft models, it was proved that quercetin nearly abolishes the expression of two DNMTs (DNMT1 and DNMT3a). Quercetin also reduced the expression of HDACs of Class I by facilitating their proteasomal degradation. Besides, *DAPK1 and BCL2L11* (pro-apoptotic genes) were demethylated by quercetin in a time- and dose-reliant fashion. Quercetin caused hyperacetylation of core histones H3 and H4 located in the promoter regions of various genes, including above two and *BNIP3, APAF1, BAX* and *BNIP3L*. Furthermore, the expression of transcripts of these genes was raised in quercetin-treated cells in comparison to solvent controls (Alvarez et al. 2018). It seems that quercetin attenuates the expression of certain DNMTs, isozymes of Class I HDACs, which in turn caused demethylation of certain genes involved in apoptosis, following which histone hyperacetylation occurred in promoters of these genes. This hyperacetylation induced the expression of these genes leading to therapeutic effect.

5.6.2 Antineoplastic Activity of Kaempferol

The broad spectrum therapeutic properties of phytochemical kaempferol have garnered the attention of scientists pandemically to concentrate on its anticancer effects. This flavonol leads to anticancer effect through multiple mechanisms. These mechanisms include antiangiogenic activity, cell cycle arrest, antiproliferative activity, antimetastatic property and induction of apoptosis. The antiproliferative action of kaempferol has been recorded in colorectal cancer, ovarian cancer, prostate and leukemia. In Caco-2 and HT-29 cell lines, kaempferol instigated cell cycle arrest through inhibition of biosynthesis of DNA (Gutiérrez del Río Menéndez, Villar & Lombó 2016).

Kaempferol was scrutinized against squamous cell carcinoma of oesophagus and the signaling mechanism modulated was also evaluated. A significant reduction in proliferation and clone formation has been seen once these cells were exposed to kaempferol. Treated cells showed induction of cell cycle arrest and dramatic change in the expression of cell cycle regulatory protein. Tumour glycolysis was also found to be the target of this flavonol and significant attenuation of this glycolysis was associated with treated cells. Hexokinase-2 down-regulation substantially reducing glucose import and lactate generation was also evident. Kaempferol inhibited the activity of epidermal growth factor receptor and suppression of its downstream signalling cascades. Exogenous up-regulation of this receptor significantly reversed the inhibitory effect of this flavonol on glycolysis, suggesting that EGFR inhibition has a crucial role in kaempferol-induced glycolysis suppression. Tumour tissue (kaempferol treated) demonstrated marked decline in EGFR activity and hexokinase-2 level (Yao et al. 2016). This supports the idea that kaempferol works both on *in vitro* and *in vivo* models of oesophageal squamous cell carcinoma.

It is established that triclosan facilitates the MCF-7 (breast cancer cells) viability through estrogen receptor α. Similarly, 17β-estradiol also promotes MCF-7 cell viability. Triclosan or 17β-estradiol-induced cell grown was considerably inhibited by kaempferol. Triclosan modulated several molecular players for promoting growth of aforesaid cells. The expression of these molecular players was reversed by kaempferol. For instance, the phosphorylation status of ERK, AKT, IRS-1 and MEK1/2 was enhanced by triclosan. On the other hand, kaempferol down-regulated the phosphorylated form of all these proteins. 17β-estradiol or triclosan-induced tumour growth in xenograft mouse model. This tumour growth was suppressed when these mice were treated with kaempferol (Kim, Hwang & Choi 2016). Again, it is clear that kaempferol works not only under *in vitro* conditions against breast cancer but also under *in vivo* conditions.

Kaempferol at low dose (20 μmol/L) showed profound suppressive effect not only on migration but also invasion of triple-negative breast cancer cells. However, no effect was observed on breast cancer cells lacking the status of being triple-negative (MCF-7 and SK-BR-3). Activation of Rac1 and RhoA was inhibited by kaempferol in the triple-negative cells. The kaempferol-induced effect in MDA-MB-231 (triple-negative breast cancer cell line) was rescued on overexpression of HER2 in these cells, indicating that the kaempferol-generated effect has strong cross-talk with HER2 (Li et al. 2017). One more study, while disclosing the effect of kaempferol on MCF-7 cells, certified that this flavonol suppresses growth of these cells through induction of programmed cell death and inhibition of Bcl2 expression (Yi et al. 2016).

Kaempferol curbed the proliferation of MDA-MB-231 breast cancer cells more effectively than the BT474 cell line (estrogen receptor positive). This effect was accompanied with G_2/M arrest, apoptosis, in addition to DNA damage in triple-negative breast cancer cells. In comparison to the control group, treated cells displayed cleavage of caspase-9 and 3, enhanced γH2AX and p-ATM (Zhu & Xue 2019). This result suggests that kaempferol is highly efficient against triple-negative breast cancer cells, which are arduous to treat. Pharmacologically beneficial effects of kaempferol were studied on breast cancer cells. Kaempferol reduced the viability of these cells in duration and concentration-dependent mode. Treatment with this flavonol provoked apoptotic death and aging by way of suppressing the PI3K/AKT apart from the human telomerase reverse transcriptase (hTERT) pathway (Kashafi et al. 2017).

Even in hepatocellular carcinoma HepG2 cells, kaempferol was authenticated to decrease cell viability, induce apoptosis concentration dependently and promote the activity of lactate dehydrogenase. Kaempferol-generated apoptotic signalling was mediated through ER stress- C/EBP homologous protein (CHOP)-signalling (Guo et al. 2016). In ovarian cancer cells, kaempferol incited cell cycle arrest through different pathways including Chk2/Cdc25C/Cdc2 and Chk2/p21/Cdc2. It was found that kaempferol-produced apoptosis and p53 induction are not reliant on Chk2. However, it was dissected that this flavonol facilitated apoptosis in these cells through the extrinsic pathway involving death receptors/FADD/caspase-8 (Gao et al. 2018; Ren et al. 2019).

In a variety of bellicose tumours, flavonol kaempferol has portrayed anticancer activity. Like other cancers, the therapeutically relevant effects of this phytochemical have been divulged in cholangiocarcinoma. Besides inhibition of cholangiocarcinoma cell proliferation, declined colony formation and stimulation of apoptosis was seen in QBC939 and HCCC9810 cells. The migration tendency and invasion capacity of these cells was markedly inhibited by kaempferol as confirmed by transwell and other assay (wound healing). While certain proteins such as Bcl-2 were down-regulated, the enhanced expression of Fas, Bax, cleaved PARP, cleaved caspase-3, 8 and 9 was quantified. Last, but no way least, p-AKT, MMP2 and TIMP2 levels were declined following the intervention with kaempferol. When tested on *in vivo* subcutaneous xenografts, kaempferol showed significant reduction of these xenografts. While the volume of subcutaneous xenograft in the control group was measured to be 0.6 cm³, this volume in the group subjected to kaempferol was only 0.15 cm³ suggesting considerable shrinkage (Qin et al. 2016).

5.6.3 ANTINEOPLASTIC CHARACTERISTICS OF FLAVONOL ISORHAMNETIN

This flavonol is one of the crucial components present in the leaves of *Ginkgo biloba* L. and in the *Hippophae rhamnoides* L. fruits. Isorhamnetin has shown a wide range of pharmacological properties. Phytochemical isorhamnetin possesses antioxidant, antitumour and antiinflammatory properties (Gong et al. 2020). Broad range antitumour activity has been attributed to isorhamnetin. It has proved beneficial against multiple cancers such as colon cancer, lung cancer, cervical cancer, gastric cancer, nasopharyngeal cancer, gastric cancer, liver cancer, pancreatic cancer and breast cancer (Yang et al. 2003; Li, Wang & Zhang 2008; Jiang, Xiang & Zhong 2012; Li et al. 2012; Antunes-Ricardo et al. 2014; Ren et al. 2015; Wang et al. 2018; Hu et al. 2019). Isorhamnetin blocks proliferation of cancer cells, triggers apoptotic death and controls proto-oncogenes and tumor suppressor genes (TSGs) (Gao, Zhang & Yu 2014).

While exploring the impact of isorhamnetin on breast cancer cells, it was noticed that this inhibitor suppresses proliferation and causes these cells to undergo apoptosis. The inhibition of a couple of signalling pathways, namely Akt and MEK, was the impetus behind this effect. The activation of these pathways through EGF blocked the isorhamnetin-induced effects, signifying that this flavonol acts through these pathways (Hu et al. 2015). Growth arrest in PANC-1 (an advanced pancreatic adenocarcinoma) cells, early apoptosis, cell cycle arrest at S-phase was prompted by isorhamnetin. These effects were connected to cyclin A down-regulation, alleviated p-MEK and ERK. Phosphorylated forms of these proteins have critical importance in cell cycle regulation, apoptosis and differentiation (Wang et al. 2018). Collectively, these results tempt me to surmise that isorhamnetin may soon prove as a propitious therapeutic.

Recently, isorhamnetin has been evaluated on two gall bladder cancer cell lines (NOZ and GBC-SD). The attenuation of cell proliferation besides G_2/M phase arrest and apoptosis induction occurred in these cells once they were subjected to the isorhamnetin environment. These effects were imputed to obstruction of PI3K/AKT signal transduction (Zhai et al. 2021). Similar results were observed in another study where isorhamnetin was tested on human bladder cancer cells. Isorhamnetin stimulated ROS production which in turn activated G2/M arrest and subsequent apoptosis. Large numbers of molecular targets were modulated by isorhamnetin in these cells. Certain proteins, such as cyclin B1, Wee1, B-cell lymphoma 2 (Bcl-2), were down-regulated while others like the cyclin-dependent kinase (Cdk) inhibitor p21[WAF1/CIP1], Fas/Fas ligand and Bcl-2 associated X protein (Bax) were up-regulated (Table 5.3). On top of that, through alteration of the mitochondrial function isorhamnetin inhibited adenosine 5'-monophosphate-activated protein kinase (Park et al. 2019).

5.6.4 Myricetin and Fisetin as Anticancer Flavonols

This polyphenol has shown different pharmacological characteristics. Substantial studies have confirmed its anticancer activity, which occurs due to its multifarious biological action. The pharmacological effect of myricetin is accredited to its ability to modulate distinct molecular targets having cross-talk with apoptosis, cell proliferation, invasion, metastasis and angiogenesis. Through induction of programmed cell death via extrinsic and intrinsic mechanisms besides reactivating silenced tumour suppressor genes, this inhibitor deters cancer advancement (Afroze et al. 2020). As myricetin possesses one hydroxal group more than quercetin it is also termed hydroxy quercetin. Myricetin occurs in glycosidically-conjugated form as well as in free form (Semwal et al. 2016). The study of epigenetic mechanisms modulated by flavonol myricetin is still at the infancy stage. Myricetin has been proved to inhibit DNMT1 activity, while certain reports suggest that its acts as an activator for non-classical

TABLE 5.3
Compendium of Various Targets Altered by Different Flavonols for Eliciting Apoptosis in Cancer Cells

Flavonol	Malignancy Name	Targets Blocked	Targets Facilitated
Quercetin	Leukaemia	HDACs	Caspase-8, c-Jun, HATs
	Breast cancer	Dehydrogenase 1A1, C-X-C, Mucin 1 p21, phospho-p38MAPK, Twist	
	Pancreatic cancer	c-Myc, TGF-β1,	TGF-β1/Smad2/3, snail1 and Zeb2
	Oral squamous cell carcinoma	Thymidylate synthase	Caspase-3,
	Oral Cancer		ROS, Fas, FADD, gastrin-releasing peptide-78, ATF-6β, (ATF)-6α, endonuclease G, cytochrome c
	Hepatocellular carcinoma	Hexokinase-2, AKT/mTOR	
	Acute myeloid leukaemia	DNMT3a, DNMT1, Class I HDACs	H3 and H4 acetylation, APAF1, BNIP3L, BNIP, BAX
Kaempferol	Oesophageal cancer	Hexokinase-2, EGFR	
	Breast cancer	p-MEK1/2, AKT,p- ERK, p-IRS-1 RhoA, Rac1, HER2, Bcl2 PI3K/AKT, hTERT	p-ATM, γH2AX,
	Hepatocellular carcinoma		Lactate dehydrogenase
	Cholangiocarcinoma	Bcl-2, p-AKT	Bax , Fas
Isorhamnetin	Breast cancer	MEK, Akt	
	Pancreatic adenocarcinoma	p-MEK, p-ERK	
	Gall bladder	PI3K/AKT, Wee1, cyclin B1, Bcl-2, AMPK	ROS, Bax, p21WAF1/CIP1

HDACs termed sirtuins (Lee, Shim & Zhu 2005; Chen et al. 2020). Fisetin, another member of flavonols, has anticancer potential. It inhibits growth of cancer cells, obstructs cell cycle progression, promotes apoptosis and modulates Bcl-2 family proteins (Imran et al. 2021). It has demonstrated anticancer potential both on cultured cells and in xenograft tumour models. Fisetin modulates cell cycle regulators and various kinases involving in cancer signalling (Rengarajan & Yaacob 2016; Fatima et al. 2021).

5.7 ANTITUMOUR ACTIVITY OF FLAVONES

Like isoflavone and flavonols, flavones come under the umbrella of flavonoids. In fact, this class of flavonoids is relatively the most well studied in the context of HDACs compared with the remaining flavonoid classes. The concept of flavones will become clearer by discussing these flavones singly with reference to cancer and HDAC enzymes.

5.7.1 APIGENIN-BASED ANTICANCER THERAPY

The off-centre expression of certain classical HDACs results not only in tumour initiation, but also strengthens progression. Apigenin has innate HDAC inhibitory strength, due to which it tunes the transcriptional dysregulation created by abnormal HDAC expression. Flavone apigenin arrests cell growth and supports apoptosis through the inhibition of classical HDACs, especially HDAC1 and HDAC3 (Pandey et al. 2012; Ganai 2017). Apigenin hampered proliferation in hepatoma cells through modulation of multiple microRNAs. This flavone collapsed tumour volume, facilitated the necrosis of tumour-constituting cells and diminished the Ki67 (proliferation marker) in xenograft tumours (Wang, S. M. et al. 2021). Human breast cancer cells portrayed declined expression of mRNA transcripts of various matrix metalloproteinases (MMP-7, 9,11,1 and MMP-3). Invasion apart from the migration of these cells also got lowered following apigenin treatment (Rajoriya 2021). Altogether, this gives an indication that apigenin has the potential to serve as a candidate molecule for attenuating breast cancer metastasis. In another eye-catching study, pancreatic cancer cell growth was inhibited by apigenin by down-regulating p-cdc2 activity and subsequent arrest of cells at G2/M. Also p-cdc25 and cyclin A and B levels were mitigated by apigenin (Ujiki et al. 2006).

Apigenin abbreviated the colon cancer cell proliferation, facilitated cleavage of PARP, stimulated apoptosis dose dependently. Bcl-xL and Mcl-1, the two antiapoptotic proteins, were repressed by the intervention of apigenin. Cell proliferation was considerably reduced and the apoptosis onset was recorded when the expression of these antiapoptotic proteins was knocked down in concert through small interfering RNA. The phosphorylation status of STAT3 known to target Mcl-1 and Bcl-xL decreased on use of apigenin (Takahashi et al. 2012). Exciting results were achieved when apigenin was tested on cervical cancer lines C33A and HeLa. Apigenin reduced cell viability and caused cell cycle blockade, prevented migration and finally epithelial to mesenchymal transition. The growth of tumour xenografts formed of C33A cells was inhibited by apigenin. This effect of apigenin was credited to its ability to suppress PI3K/AKT and FAK signalling pathways, due to which certain molecular targets such as Bax, CDC25c, CDK1, Bcl-2, N-cadherin, E-cadherin, laminin and vimentin were modulated, thereby yielding pharmacological benefits (Chen et al. 2022).

Apigenin reduced viability and induced apoptotic form of cytotoxicity in DU145 and PC-3 cells which are androgen-refractory. Apigenin diminished the expression of various proteins considered as inhibitors of apoptosis, such as c-IAP1, XIAP and survivin. Escalation of cytochrome C, Bax along with down-regulation of Bcl-2 and in

Bcl-xL was quite conspicuous after these cells were given apigenin as per the proper procedure. The dissociation of Bax from Ku70 is critical for inducing apoptotic signalling. Apigenin inhibited Class I HDAC isozymes and declined the protein levels of HDAC1, thereby augmenting Ku70 acetylation. This acetylation serves the purpose of Ku70-Bax dissociation and ultimately apoptosis of cancer cells. The occupancy of HDAC1 at the promoter region of XIAP post apigenin exposure induces the downregulation of the latter. Apigenin demonstrated its apoptotic effect even in athymic nude mice xenograft models, suggesting that it works even in three-dimensional disease models (Shukla, Fu & Gupta 2014). Collectively, apigenin targets HDACs for disruption of Bax-Ku70 interaction and thus follows epigenetic mechanism for inducing apoptosis in androgen-refractory cancer cells and tumours.

IKKα has a strong role in metastasis and is thus regarded as a candidate molecular target for developing anticancer drugs. Apigenin inhibits the kinase activity of IKKα at lower concentration (micromolar). It exerts antiinvasive and antiproliferative effects on prostate cancer cells through attenuation of the kinase activity of the above-mentioned kinase besides bridling the NF-κB/p65 activation. Apigenin proved relatively better than PS1145 (IKK inhibitor) in suppressing this cancer-promoting kinase. Apigenin-induced cell cycle arrest in prostate cancer cells resembled that of IKKα knockdown (Shukla et al. 2015; Ji et al. 2022). Mitochondrial adenine nucleotide translocase (ANT) promotes ADP-ATP exchange across the mitochondrial membrane (inner one) and thus has a key role in energy metabolism. Among its four forms, ANT2 resides majorly on undifferentiated, proliferative cells and imports the ATP generated through glycolysis to mitochondria. ANT2, unlike ANT3 and ANT1, is not pro-apoptotic and its expression serves a sign for carcinogenesis (Chevrollier et al. 2011). Apoptosis induced by Apo2L/TRAIL was enhanced by genistein and apigenin in cancer lines. While apigenin facilitated the expression of DR5, genistein failed to do so. Also, unlike apigenin, ANT2 was not a target of genistein. ANT2 knockdown produced effects similar to apigenin intervention including DR5 up-regulation and increased Apo2L/TRAIL-driven apoptosis (Oishi et al. 2013). This establishes that apigenin inhibits ANT2 and thus elevates the expression of DR5, thereby facilitating Apo2L/TRAIL-induced cytotoxic effects.

The last enzyme in glycolysis, namely pyruvate kinase, transforms phosphoenolpyruvate to pyruvate results in the generation of ATP through substrate level phosphorylation. Among the four isoforms of this enzyme, PKM2 is broadly expressed in multiple tumour types and has connection with tumorigenesis (Zahra et al. 2020). Apart from its pyruvate kinase function, this enzyme performs a nonmetabolic activity in cancers (Wong, De Melo & Tang 2013). Once this kinase undergoes nuclear translocation, it carries out phosphorylation (protein kinase activity) of a variety of proteins and thus facilitates physiopathological events (Chen, Chen & Yu 2020). Substantial anticolon cancer effect has been evinced while using apigenin on colon cancer cells. Apigenin obstructed glycolysis in these cells by inhibiting activity and also expression of tumour-specific pyruvate kinase M2. It has been proved in HCT116 cells that apigenin is an allosteric inhibitor of this enzyme as D-fructose-1,6-diphosphate (FBP) was not able to reverse the restraining effect of apigenin on PKM2 (Shan et al. 2017).

Apigenin was studied against osteosarcoma with the aim to decipher the underlying molecular mechanism elicited. Osteosarcoma is nothing but a bone tumour that is malignant. It occurs both in adolescents and children. A range of concentrations of apigenin were investigated on the SOSP-9607 osteosarcoma cell line. The influence of this flavone on proliferation, invasion, migration, Warburg effect and stemness was assessed. Apigenin showed multiple effects on this cell line. It reduced proliferation and obstructed epithelial to mesenchymal transition of these cells on the one hand and on the other hand restrained their migration and invasion. The expression profile of E-cadherin was enhanced while that of vimentin was lowered. Apigenin declined the number of cells possessing stem cell markers such as OCT-4 and Nanog. The Warburg effect in these cells was also inhibited by apigenin, as indicated by the reduced levels of glucose and lactic acid, elevated ATP and citrate levels. The down-regulation of molecular players linked to the Warburg effect, including LDHA, HK1 and GLUT1, was also prominent in apigenin-encountering cells (Shi, Lian & Jia 2022). Phosphorylated forms of various proteins encompassing PI3K, mTOR and Akt were also lowered, providing clues that apigenin-induced effects are transmitted through the PI#K/Akt/mTOR pathway.

5.7.2 CHRYSIN AS PROPITIOUS MOLECULE FOR BELLIGERENT CANCERS

This hydroxylated flavone mainly occurs in propolis, *Passiflora incarnate*, *Pelargonium crispum*, *Oroxylum indicum* (Manzolli et al. 2015; Mehdi et al. 2018). Like apigenin, chrysin has manifested activity against a wide range of tumours. It has demonstrated activity against breast cancer, melanoma, cervical cancer, hepatocellular carcinoma and colorectal cancer (Ganai, Sheikh & Baba 2021). Growth of MCF-7 breast cancer cells was curbed by chrysin in a time- and dose-dependent form. This inhibition of growth has been credited to its ability to induce apoptosis in these cells (Samarghandian et al. 2016). Following an intrinsic apoptotic pathway, chrysin transmits apoptotic signalling in bladder cancer cells. This is obvious from the activation of caspse-3 and 9 and not caspase-8. Chrysin treatment was associated with pacified expression of certain antiapoptotic proteins covering Mcl-1, Bcl-2 and Bcl-xl but increased Bcl-2 associated X expression. Chrysin triggered endoplasmic reticulum stress and ROS production in these cells and blocked the STAT3 pathway (Xu et al. 2018). N-acetylcysteine use reversed the chrysin-induced effect in these cells, signifying that ROS has a key role in chrysin-transmitted therapeutic effect.

Chrysin promotes cytostatic behaviour in HeLa cells and attenuates their migration potential. The methylation status of multiple tumour suppressor genes was lessened by chrysin, due to which their expression increased both at transcript and protein levels. Furthermore, the degree of expression of a variety of epigenetic-modifications modifying enzymes including HATs, HDACs, HMTs and DNMTs was lowered post-chrysin incubation. Chrysin reduced the DNA methylation all through the genome and modulated post-translational marks on histones H3 and H4 (Raina et al. 2022). This reveals that chrysin targets chromatin-modulating enzymes for hampering the migration strength of human cervical cancer cells. Chrysin alters the stability of genome in breast cancer cells, prevents their survival and makes them more sensitive to chemotherapeutic agents. Chrysin-facilitated genomic instability has been connected to its potential to impede repairment of DNA double-strand breaks culminating in DNA damage amassing (Geng et al. 2022). In murine hepatic carcinoma cells, chrysin impels antitumorigenic effects. Chrysin by way of down-regulating programmed death ligand 1 blocked the progression of hepatocellular carcinoma. This was noticed both under conditions of *in vitro* and in xenograft models (Rong et al. 2022).

5.7.3 SANGUINE ANTINEOPLASTIC EFFECTS OF LUTEOLIN

Like apigenin and chrysin, luteolin has manifested considerable antineoplastic activity against a myriad of cancer types. However, unlike them luteolin is extensively studied, especially with reference to its modulation of epigenetic signalling mechanisms (Ganai et al. 2021). The anticancer effect of luteolin on breast cancer cells was experimentally verified on MDA-MB-231 cells. Apart from exerting antiproliferative effect, this flavone increased the speed of apoptosis. Luteolin altered the expression of various targets, namely human telomerase reverse transcriptase (hTERT), c-Myc and nuclear factor-κB inhibitor α (Huang, Jin & Lan 2019). Another studied linked the antiproliferative activity of luteolin to induction of target genes including *p27* and *p21* of FOXO3a (Lin et al. 2015). While examining the impact of luteolin on breast tumours (estrogen receptor negative), it was concluded that luteolin induces tumour growth inhibition. On delineating the molecular mechanism it was found that luteolin down-regulates the transcripts of EGFR and thus hampers MAPK activation reliant on the former (Lee, Oh & Sung 2012). One more study exploring the therapeutic effects of luteolin revealed its antiproliferative and antimetastatic activity on triple-negative breast cancer cells (androgen receptor positive). Epithelial to mesenchymal transition is a key event in metastasis. Luteolin reversed this transition by suppressing the AKT/mTOR signalling and subsequent inhibition of MMP9. Besides, MMP-9 down-regulation was associated with the declined acetylation status of histone H3 lysine 27 (H3K27) and histone H3 lysine 56 (H3K56) in its promoter area (Wu et al. 2021). A current study also favoured the proliferation inhibition properties of luteolin. In a couple of breast cancer lines, luteolin attenuated the proliferation and restrained motility by regulating the expression of many oncogenes (Wang et al. 2022). Luteolin suppressed invasion and growth of gastric cancer cells by modulating multiple players. Luteolin decreased the expression of Notch1 and p-mTOR, p-ERK, p-STAT3. While the expression of miRNAs functioning as tumour suppressors such as miR-422a, miR-107, miR-34a and miR-139) was increased, the expression status of oncogenic miRNAs covering miR-224, miR-340, miR-155 and miR-21 was significantly lowered (Pu et al. 2018). Luteolin blocked the phosphorylation of STAT3 in gastric cancer cells having overactivated STAT3. The inhibition of STAT3 phosphorylation in turn suppressed the expression of its target genes, namely Bcl-xl, Survivin and Mcl-1. Interestingly, the luteolin-induced effects on STAT3 and cell cytotoxicity were obliterated on silencing the protein tyrosine phosphatase SHP-1, signifying the key role of this phosphatase in mediating luteolin signalling. In complex conditions with HSP90, the dephosphorylation of STAT3 by the above-mentioned phosphatase is shielded. Luteolin breaks the interaction of STAT3 with HSP90, making the ground smooth for SHP-1 to dephosphorylate STAT3. The growth inhibitory activity of luteolin was also recorded in the three-dimensional xenograft mouse model carrying the GC tumour (Song et al. 2017). The bottom line is that luteolin induces the segregation of HSP90 and STAT3, making the latter prone to SHP-1 dephosphorylation, culminating in reduction of STAT3 activity. The expression of miR-34a is comparatively lower in gastric cancer tissues than non-tumour tissue (pair matched) and this miRNA negatively regulates Bcl-2. Proliferation inhibition and the invoking of apoptosis was manifested in gastric cancer cells after they were subjected to luteolin addition. By way of enhancing the expression of miR-34a, luteolin lowers the protein level of Bcl-2, thereby inducing apoptosis in gastric cancer cells. The luteolin-induced inhibition of Bcl-2 was partially rescued by anti-miR-34a oligonucleotides, suggesting that some other pathway is also involved in cooperative down-regulation of Bcl-2 (Wu et al. 2015).

In cervical cancer cells positive for human papilloma virus (HPV), luteolin induced statistically substantial cytotoxicity in comparison to cervical cancer cells negative for HPV (C33A). Oncogenes of this virus, namely E6 and E7, were inhibited by luteolin. Flavone luteolin promoted activation of caspases 3 and 8 dose dependently, collapsed mitochondrial transmembrane potential, and increased cytosolic cytochrome c, besides pacifying expression of Bcl-2 and Bcl-xL (Ham et al. 2014). Time- and dose-dependent cytotoxic effect was seen in HeLa cells following luteolin exposure. While the antiapoptotic genes such as *BCL-2*, *MCL-1* and *NAIP* underwent down-regulation, pro-apoptotic genes including *FAS*, *TRADD*, *BAX*, *BOK*, *APAF1*, *BAK1* and *BID* exhibited a reverse trend (Raina et al. 2021). Myocyte enhancer factor 2D (MEF2D) has severe implications in hepatocellular carcinoma advancement. It favours progression of hepatocellular carcinoma through AMOTL2/YAP signalling. Basically, MEF2D activates YAP signalling by modulating AMOTL2, the negative regulator of aforesaid signalling. Luteolin inhibited the activation of this pathway thereby attenuating the progression of defined cells (Xu et al. 2022). Luteolin was recently tested on clear cell renal cell carcinoma cells for its cytotoxic effect. The outcome from the study is that luteolin significantly reduced the survival of these cells. This was the case both under *in vitro* and *in vivo* conditions and was associated with atypical GSH depletion and elevated cellular Fe^{2+}. Thus luteolin induced ferroptosis in these cells, as these effects were rescued to certain levels by deferiprone (iron-ion chelator) (Han et al. 2022). The therapeutically relevant effects of luteolin were observed on colorectal cancer cells. These cells after treatment with the said inhibitor demonstrated reduction in proliferation and cell survival, cell cycle arrest, DNA damage and apoptotic death. Luteolin intervention resulted in up-regulation of p-CHK1, while down-regulated the levels of antiapoptotic players. Luteolin inhibited the MAPK signalling in predefined cells, indicating that the effect of the former is intervened through the MAPK pathway (Song et al. 2022).

The pharmacological effects of luteolin on melanoma models have been explored. The viability of melanoma cells underwent decline post-incubation with luteolin. The inhibition of invasion, migration and programmed cell death induction occurred in B16F10 and A375 cells. Src is a kinase that functions upstream of STAT3. Luteolin obstructed phosphorylation of both these proteins, promoted STAT3 clearance and inhibited its downstream target genes. Mice models bearing melanoma showed a collapse of melanoma growth and inhibition of Src/STAT3 signaling after preventive delivery of luteolin. In A375 cells, STAT3 hyper-activation weakened the antimelanoma activity of luteolin (Li et al. 2022). In single line it can be concluded that luteolin has promising antimelanoma effect that is connected to its ability to target STAT3. Migration of HeLa cells and their colony-forming ability was curbed by luteolin. Luteolin modulated a variety of epigenetic enzymes including HDACs, histone methyltransferases, DNA methyltransferases and histone acetyltransferases and declined methylation of DNA globally. Decline in methylation status at the promoters of tumor suppressor genes reactivated their expression. Among these genes *TP53*, *TIMPS*, *CDH1*, *FHIT*, *PTEN*, *TP73* and *DAPK1* were conspicuous (Pramodh et al. 2022). The impact of luteolin was inspected in vascular endothelial cells of non-small cell lung cancer (NSCLC-VECs). The angiogenesis, migration, invasion and viability of these cells was hampered by this marvellous flavone. Post-incubation with luteolin escalated miR-133a-3p and inhibitor of the latter thwarted luteolin-engendered effects. Proteins facilitating migration and invasion including MMP2/9, MAPK and PI3K/Akt were inhibited after the addition of luteolin to these cells (Pan et al. 2022).

5.7.4 RUTIN AS ANTICANCER FLAVONE

In spite of possessing low nutritional value, flavones offer a high degree of therapeutic benefits. Rutin is an emerging flavone in the context of anticancer activity. This flavone has displayed a promising effect on human glioma (CHME) cells. MTT assays proved that rutin triggers high-level cytotoxic effect in these cells by inducing up-regulation of p53. The cytotoxic effect was apoptotic as nuclear condensation followed by fragmentation and budding bodies were confirmed in treated cells. Rutin enhanced the production of reactive oxygen species and promoted mitochondrial membrane depolarization, thereby invoking the intrinsic pathway of apoptotic death in these cells (Yan et al. 2019). The knockdown of P53 counteracted rutin-signalled effects, suggesting that the effects of this flavone are interceded by this tumour suppressor. While scrutinizing the impact of rutin on neuroblastoma cell line namely LAN-5, it was established that this flavone inhibits chemotactic capacity and growth of this cell line. Also, cell cycle arrest and apoptotic death was evident after rutin application. The ratio of BCL2/BAX was lowered along with transcript levels of MYCN and TNF-α secretion (Table 5.4) (Chen et al. 2013). These results support the preclinical antineuroblastoma effect of rutin. *Abrus precatorius*, a medicinal plant of the tropics, possess seeds whose antiproliferative and antioxidant characteristics have been examined on HeLa and Hep2C cells. The seed extract of this plant contains rutin as its one of the constituents, the other being tannic acid. The extract obtained through different methods has shown antiproliferative activity against these cell lines to a different extent (Kaur et al. 2022). Thus, the bioactive molecules present in the seed extract of the above-mentioned plant have antioxidant and anticancer effect against cervical cancer cells. Rutin was evaluated for its anticancer property on a couple of breast cancer cell lines and a single pancreatic cancer model. Rutin manifested anticancer activity *in vivo*, as it resulted in the collapse of tumour dimensions. Rutin exerted cytotoxicity on Panc-1 and MCF-7 lines by facilitating apoptotic cell death (Saleh et al. 2019).

TABLE 5.4
Mechanism Followed by Flavone Histone Deacetylase Inhibitors for Apoptotic Signalling in Malignant Cells

Flavone	Abnormality name	Activated molecules	Inhibited molecules
Apigenin	Hepatoma		Ki67
	Pancreatic cancer	cyclin A, cyclin B	p-cdc2
	Colon cancer		Mcl-1, Bcl-xL p-STAT3
	Cervical cancer		FAK, PI3K/AKT
	Prostate cancer	cytochrome C, Bax	XIAP, c-IAP1, survivin, Bcl-2, Bcl-xL, HDAC1
		DR5	IKKα, ANT2
	Colon cancer		Pyruvate kinase M2
	Osteosarcoma	E-cadherin	Vimentin, LDHA, HK1, GLUT1
Chrysin	Breast cancer	Caspse-3/9, Bax	Bcl-xL, Bcl-2, STAT3
	Cervical cancer		HDACs, DNMTs, HMTs
	Hepatic carcinoma		PDL1
Luteolin	Breast tumour		EGFR, MAPK
	Triple-negative breast cancer		AKT/mTOR, MMP9, H3K27ac, H3K56ac
	Gastric cancer	miR-422a, miR-34a, miR-107, miR-139	Notch1, p-ERK, p-mTOR, p-STAT3
	Cervical cancer		Bcl-xL, Bcl-2
		TRADD, APAF1, BID, BOK	MCL-1, BCL-2, NAIP
	Colorectal cancer	p-CHK1	MAPK
	Melanoma		p-src and p-STAT3
	Cervical cancer	CDH1, TP53, FHIT, TP73, DAPK1	DNA methylation
	Non-small cell lung cancer	miR-133a-3p	PI3K/Akt, MAPK, MMP2/9
Rutin	Glioma	p53, ROS,	
	Neuroblastoma		BCL2/BAX, MYCN

5.8 FLAVANOLS AND THEIR ROLE IN CANCER THERAPY

This class of flavonoids occurs mostly in fruits, tea, cereals and cocoa. They are nearly absent in legumes and vegetables, with the exception of broad beans and lentils. They also occur in peels of fruits or their seeds (de Pascual-Teresa, Santos-Buelga & Rivas-Gonzalo 2000; Määttä-Riihinen et al. 2005). Epigallocatechin gallate (EGCG), the flavanol from green tea, exhibits a broad range of biological functions. Catechin has been endorsed as an inhibitor of DNMTs. EGCC enhances the site-specific acetylation of histone H3 and also increases site-specific methylation marks (H3K9me3 and H3K4me3). Catechin has also demonstrated HDAC inhibitory activity in cellular and cell free conditions. EGCG modulates chromatin topology by suppressing the expression of proteins involved in the propagation of heterochromatinization, and this favours chromatin decondensation in endothelial cells of human beings (Ciesielski, Biesiekierska & Balcerczyk 2020). This flavanol inhibited growth of colon cancer cells, tumour cell proliferation and microvessel density by 58%, 27% and 30% respectively. EGCG elevated apoptosis of these cells by 1.9 fold and that of endothelial cells by 3 fold in comparison to control. Flavanol EGCG partly exerts anticancer activity by blocking VEGF expression, which in turn restrains angiogenesis (Jung et al. 2001). An attempt has been made where it has been addressed whether EGCG follows epigenetic route in regulating the expression of Raf kinase inhibitor protein (RKIP) and the metastatic potential of pancreatic adeno-carcinoma cells (AsPC-1). Pharmacological intervention with EGCG elicited the up-regulation of RKIP by way of restraining HDAC function. The expression of multiple molecular players encompassing MMP-2/9, Snail, nuclear shuttling of NF-κB was inhibited by this flavanol in the aforesaid cell line. While the ERK phosphorylation was mitigated, the E-cadherin expression underwent up-regulation in EGCG-treated cells (Kim & Kim 2013). COX-2 overexpression is involved in a variety of pathologies and also in cancer. Inhibition of this enzyme has been surmised to prove an effective strategy to overcome cancer. Selective inhibitors of this enzyme have proved significantly effective against some cancer types. EGCG, the premier component of green tea, has inherent anticancer potential. This flavanol inhibits COX-2 and leaves COX-1 unaltered in androgen insensitive and androgen-sensitive prostate cancer cells (Hussain et al. 2005; Almatroodi et al. 2020).

Catechin has been investigated for its therapeutic potential against non-small cell lung cancer cells. At 600 μmol·L^{-1}, 19.76% inhibition of A549 cell proliferation was quantified after 24 hours duration. This effect of catechin was attributed to its ability of elevating the expression of p21 in these cells. Moreover, p-AKT and cyclin E1 levels were suppressed in treated cells following dose-dependent trend (Sun at al. 2020). Catechin hydrate induced apoptosis in breast cancer (MCF-7) cells as revealed by contemporary TUNEL (terminal deoxynucleotidyl transferase-mediated dUTP nick end labelling) assay. Better cytotoxic effect was achieved after 48 hours duration. While catechin hydrate at 150 μg/ml induced apoptosis in 40.7% cells after 24 hours, the same concentration induced 43.73% apoptosis when plates were kept for 48 hours incubation. Similarly, catechin hydrate when given at concentration of 300 μg/mL resulted in 41.16% and 52.95% apoptosis at 24 and 48 hours respectively. Both catechin hydrate concentrations when used for 72 hours were equally effective in inducing loss of A549 cells integrity. Catechin hydrate-generated apoptosis was accompanied with elevated expression of several players such as caspase-3/8/9 and *TP53* (Table 5.6) (Alshatwi 2010). The tumour growth inhibitory effect of epicatechin (EC) was studied on triple-negative breast cancer. This type of breast cancer is peculiar as such cells lack the estrogen receptor, progesterone receptor and epidermal growth factor receptor 2. This flavanol inhibited tumour growth nearly as better as doxorubicin. While the former inhibited the defined growth by 74%, the latter inhibited by 79%. EC modulated various pathways including AMPK, phosphorylation of AKT in addition to mTOR expression culminating in obstruction of cell proliferation. EC-treated animals manifested increased survival over vehicle-treated animals (Almaguer et al. 2021; Table 5.5).

TABLE 5.5
Critical Cellular Targets Customized by Flavanols for Engendering Cancer Cell Apoptotic Death

Flavanol	Name of Malignancy	Molecules Alleviated	Molecules Heightened
Epigallocatechin gallate	Colon cancer	VEGF	
	Pancreatic adenocarcinoma	HDAC, MMP-2/9, Snail, p-ERK	RKIP, E-cadherin
	Prostate cancer	COX-2	
Catechin	NSCLC	p-AKT, cyclin E1	p21
	Lung cancer		caspase-3/8/9, TP53

5.9 ANTICANCER CHARACTERISTICS OF CHALCONES

Chalcones serve the purpose of lead molecules for designing novel therapeutics against cancer. They act as point of departure for developing new drugs with strong potency against molecular targets (McCluskey & Russell 2021). Anticancer activity of famous natural chalcones will be discussed in subsequent paragraphs, which will further enahnce the concept of chalcones and their emerging therapeutic benefits against aggressive malignancies.

5.9.1 OVALITENIN A AND ITS ROLE IN HAMPERING CANCER

Chalcone (Ovalitenin A) derived from a plant, namely *Millettia pulchra*, has been assessed on HeLa cells, and the signalling mechanism induced by it has also been delineated. This chalcone bridled proliferation of these cells and stimulated apoptotic type of death. Together with these events, morphological transformation, external-ization of phosphatidylserine and caspase-3 activation was confirmed. Ovalitenin A, apart from causing cell cycle arrest, increased the ratio of Bax/Bcl-2. Last, but no way least, this plant separated chalcone reduced the COX-2 protein level and invoked depolarization of mitochondrial membrane (Liu et al. 2016).

5.9.2 LICOCHALCONE A AS ANTICANCER CHALCONE

Licochalcone A, a chalconoid separated from the various species of *Glycyrrhiza* has demonstrated antimicrobial, antiinflammatory, antitumor, antioxidant and antiparasitic activity. This chalconoid inhibited growth of colorectal and neuroblastoma cell lines. Hypoxia induces the expression of the tropomyosin receptor kinase B (*TrkB*) gene in cancer cells. Licochalcone A inhibited hypoxia elicited expression of *TrkB* at transcript level. The brain-derived neurotrophic factor serves as ligand having high affinity against TrkB. Treatment with licochalcone A also inhibited protein kinase B activation under conditions of hypoxia (Arita et al. 2020).

In breast cancer cells, licochalcone A inhibited proliferation. This plant deriva-tive facilitated caspase-3 and 9 cleavages, suppressed expression of Bcl-2 triggering cytochrome c release. This molecule reduced proliferation, enhanced generation of ROS and promoted apoptosis. Besides, cell migration and invasion were also inhibited by this promising chalcone (Huang et al. 2019). The IFN-γ-promoted expression of PD-L1 in lung cancer cells was restrained by licochalcone A. This polyphenol compound increased ROS production in lung cancer cells, which was reversed by N-acetyl-L-cysteine (Yuan et al. 2021). Through activation of JNK/p38 in naso-pharyngeal carcinoma cells, licochalcone A induced cytotoxic effect (Chuang et al. 2019). In the cells of bladder cancer this flavonoid blocked proliferation via ROS-backed cell cycle arrest (G2/M) and apoptosis (Hong et al. 2019). Studies on two lung cancer lines including H460 unfolded that licochalcone A reduces their viability through induction of apoptosis. Treatment with defined compound modulated expression of several proteins, including MDM2, Cdc2 and Cyclin B1. Intervention was also accompanied with PARP breakage, caspase-3 activation and enhanced expression of endoplasmic reticulum stress markers ATF4 and p-EIF2α (Qiu et al. 2017). Licochalcone A exerted profound anticancer effect on non-small cell lung cancer cells by hampering EGFR signalling and by down-regulating the survivin at protein level. These results suggest that targetting survivin may be an attractive strategy to make non-small cell lung cancer cells more sensitive to conventional therapeutics (Constantinescu & Lungu 2021; Gao et al. 2021).

5.9.3 PANDURATIN A AND ITS ANTICANCER BENEFITS

Panduratin A, derived from the *Boesenbergia pandulata*, has also been studied in a cancer context. This cyclohexanylchalcone has shown selective killing of pancreatic cancer cells (Nguyen et al. 2017). Cancer cells monotonous to apoptotic signalling drugs can be killed through modulation of autophagy. Autophagy has cross-talk with prosurvival mechanism that induces chemoresistance in cancer cells. Panduratin A takes the benefit of ER-stress intercession for inducing cytotoxicity in melanoma cells. Up-regulation of LC3B, a marker of autophagy, has been quantified in melanoma cells exposed to this flavonoid. Panduratin A when used on melanoma cells evoked autophagy via suppression of mTOR signalling and by facilitating AMPK pathway. Apoptosis induced by panduratin A was heightened when an autophagy inhibitor was used in conjunction with the former (Lai, Mustafa & Wong 2018). This outcome indicates that autophagy promoted by panduratin A gives protection to melanoma cells. Growth inhibition was noticed in MCF-7 cells were given panduratin A while either little or no effect was exerted on normal breast cells MCF-10A. Lot of changes occurred in chalcone-treated cells, which together contributed towards cytotoxic effect. Among these changes, Bax up-regulation, cytochrome c release, activation of caspases 7-9 and inhibition of Bcl-2 were profound (Liu et al. 2018).

5.9.4 CARDAMONIN AS PROMISING ANTINEOPLASTIC NATURAL CHALCONE

Another chalcone, cardamonin, isolated from *Campomanesia adamantium* belonging to the Myrtaceae family, inhibited NF-κB activity in prostate cancer cells and induced fragmentation of nuclear DNA (Pascoal et al. 2014). This chalcone blocked proliferation of oesophageal cancer cells, inhibited their migration and invasion, finally causing apoptotic killing. Cardamonin prevented epithelial to mesenchymal transition and thus inhibited metastasis. This was accompanied with up-regulation of E-cadherin, an epithelial marker and down-regulation of vimentin and N-cadherin, the mesenchymal markers. Matrix metalloproteinases and Snail (EMT transcription factor) were also prominent among the down-regulated proteins. The aforesaid chalcone induced apoptosis in these cells by attenuating the PI3K/AKT signalling route (Wang, S. M. et al. 2021). A variety of edible plants, including *Alpina* species and cardamom contain cardamonin. It has shown significant anticancer property against prostate, lung, colorectal and breast cancer lines. Recently, its antiproliferative and cytotoxic activity has been demonstrated against hepatocellular carcinoma cells. Further, this natural chalcone induced G1 arrest in these cells and culminated in apoptosis through intrinsic and extrinsic routes (Badroon, Abdul Majid & Alshawsh 2020).

5.9.5 LONCHOCARPIN AND COMPENDIUM OF ITS ANTITUMOUR POTENTIAL

This natural chalcone is isolated from the plant *Lonchocarpus sericeus* and has shown anticancer character against leukaemia and neuroblastoma cells. The key event in the advancement of colorectal cancer is dysregulation of the famous Wnt/β-catenin signalling. Many flavonoids block this signalling and thus soothe inflammation, cancer onset, progression and death of neoplastic cells. *In vitro* and animal-based studies have revealed the inhibitory activity of lonchocarpin on the above-mentioned signalling. Nuclear localization of β-catenin is vitiated by this chalcone. Lonchocarpin inhibited proliferation and migration of multiple colorectal cancer cell lines without imparting overt effects on normal intestinal cell line. Tumour proliferation was also retarded in the mice model of colorectal cancer. This all clearly explains that lonchocarpin reduces colorectal cancer cell growth not only under *in vitro* environment but also under complex *in vivo* conditions (Predes et al. 2019). This beneficial effect of lonchocarpin has been accredited to its ability to cripple Wnt/β-catenin signalling.

Lonchocarpin has been evaluated using a panel of lung cancer lines, and against H292 cell line 97.5% activity was estimated. A molecular docking study disclosed that this chalcone binds to BH3-binding groove located on Bcl-2 protein through hydrophobic contact with Ala 146. A substantial decline in cell proliferation accompanied with Bax, caspase-9 and caspase-3 activation occurred due to lonchocarpin intervention (Table 5.6). While 47.9% apoptosis was recorded after 48 hours, only 41.1% death was quantified after 24 hours. Maximum inhibition on tumour growth (72.51%) in the mice model was achieved at 100mg/kg of this chalcone (Chen et al. 2017; Almatroodi et al. 2020).

5.10   ANTHOCYANINS AS ANTICANCER FLAVONOIDS

Anthocyanins are prevalent across the plant kingdom. They have shown multiple therapeutic effects against cancer lines and tumour models. The intake of anthocyanins has been linked to the reduced risk of arthritis, diabetes and cancer due to their antiinflammatory and antioxidant power (Prior & Wu 2006). In acidic solution

TABLE 5.6
Mechanism Followed by Chalcones and Anthocyanins to Actuate Death Signalling in Different Models of Cancer

Chalcone	Cancer	Downregulated players	Players upregulated
Ovalitenin A	Cervical	Bcl-2, COX-2	caspase-3, Bax
Licochalcone A	Colorectal, neuroblastoma	TrkB, protein kinase B	
	Breast	Bcl-2, PD-L1	
	Nasopharyngeal		JNK/p38
	Lung		ATF4, p-EIF2α, caspase-3
	NSCLC	EGFR, survivin	
Panduratin A	Pancreatic		LC3B
	Melanoma	mTOR	AMPK
	Breast	Bcl-2	Bax, caspases 7-9
Cardamonin	Prostate	NF-κB	
	Oesophageal	Vimentin, N-cadherin, Snail, PI3K/AKT	E-cadherin
Lonchocarpin	Colorectal	Wnt/β-catenin	

they have positive charge unlike other flavonoids (Mazza 1995). Anthocyanins exist as glycosides and once an anthocyanin is hydrolysed the non-sugar part that is left is known as anthocyanidin. Thus, anthocyanidins are aglycones of anthocyanins. While anthocyanins occur as acylated anthocyanins and anthocyanidin glycosides, anthocyanidins exist as 3-deoxyanthocyanidins, 3-hydroxyanthocyanidins and O-methylated anthocyanidins. Delphinidin, cyaniding, peonidin, pelargonidin, malvidin and petunidin are the most common anthocyanidin types. Acylated anthocyanins may be coumaroylated anthocyanins, acrylated anthocyanins, malonylated anthocyanins or caffeoylated anthocyanins (Khoo et al. 2017).

Anthocyanins derived from *Lonicera edulis* exerted antiproliferative effect when used on HT29 cells. Changes in morphology of these cells peculiar to apoptotic cells were seen. The effects were more severe when higher concentrations of anthocyanins were used (WenXing, YiHong & ZhenYu 2011). Anthocyanins were studied on HER2 positive cells as well as on trastuzumab-resistant cells. Peonidin-3-glucoside and cyanidin-3-glucoside obstructed HER2 phsophorylation, inhibited migration, invasion, induced apoptosis and shrank the growth of tumours (Li et al. 2016). One more study concluded that *Vitis coignetiae Pulliat* isolated anthocyanins decrease resistance of breast cancer cells to cisplatin through inhibition of NF-κB Activation and Akt (Paramanantham et al. 2020). The impact of peonidin 3-glucoside was studied on lung cancer cells. This anthocyanin substantially inhibited invasion, urokinase-type plasminogen activator (u-PA) and secretion of MMPs (MMP2, MMP-9) in these cells. Phosphorylation of (ERK) 1/2 regulating the expression of u-Pa and MMPs was hampered by this glucoside (Ho et al. 2010). Mice (azoxymethane/dextran sodium sulfate treated) fed with black raspberry anthocyanins demonstrated a decline in colon carcinogenesis. These anthocyanins enhanced expression of SFRP2 by inducing demethylation at the promoter regions of the gene encoding this protein. In the above-mentioned mice, black raspberry anthocyanins induced down-regulation of DNMT3b, DNMT31 and phosphorylated form of STAT3 (Chen et al. 2018).

The mechanism of tumor cell proliferation inhibition was studied using mouse colon cancer cells as models in the context of glycosylation of anthocyanins. It was observed that glycosides of anthocyanins alter glucose transport, resulting in energy metabolism inhibition, culminating in mitochondrial damage and subsequent apoptosis of tumour cells (Jing et al. 2020). The vascular endothelial growth factor receptor 2 (VEGFR2) and the epidermal growth factor receptor (EGFR) are two candidate molecular targets for circumventing non-small cell lung cancer (NSCLC). A study on these cells overexpressing EGFR/VEGFR2 revealed that anthocyanidin delphinidin potentially inhibits both these growth factor receptors. Delphinidin (5-60 μM) blocked not only the PI3K activation but also inhibited the phosphorylation of MAPKs and AKT. Delphinidin inhibited growth of NSCLC cells without imparting substantial toxicity on human bronchial epithelial cells. Delphinidin intervention on cells (SK-MES-1 and NCI-H441) induced PARP protein cleavage, caspase-3 and 9 activation, besides the down-regulation of Mcl-1, Bcl2 and Bcl-xL. Additionally, certain proapoptotic proteins including Bak and Bax underwent up-regulation on exposure to delphinidin (Pal et al. 2013). Epithelial to mesenchymal transition of breast cancer cells possessing HER2 was suppressed by anthocyanins isolated from black rice under conditions of *in vitro* through FAK signalling inhibition (Zhou et al. 2017). Shrinkage of transplanted tumour growth and suppression of pulmonary metastasis was noticed in BALB/c nude mice when they were administered with black rice derived anthocyanins at concentration of 150 mg/kg/day. These mice were carrying xenografts made of breast cancer cells that were ErbB2 positive. These anthocyanins inhibited the adhesion, migration, invasion and motility potential of these cells. The above-mentioned effects were associated with inhibition of u-PA, the metastasis-facilitating serine protease (Luo et al. 2014; Mahmood, Mihalcioiu & Rabbani 2018).

Glioma cells are well known for resistance to certain agents known as alkylating agents. Cyanidin-3-O-glucoside has the potential to restrain the growth of tumour cells. The temozolomide-resistant glioma cell line was generated (LN-18/TR) and it was found that MGMT and β-catenin is substantially up-regulated in this line. Cyanidin-3-O-glucoside through miR-214-5p up-regulation increased the temozolomide-induced cytotoxicity in LN-18/TR cells. Also, this glucoside inhibited MGMT and β-catenin (Table 5.7). Cyanidin-3-O-glucoside effects were reversed by the inhibitor of miR-214-5p, confirming that the former induces its effects through up-regulation of the latter (Zhou et al. 2022).

Up to this stage, different types of phenols have been discussed extensively. Especially, the broad classification of flavonoids, which like tannins come under the blanket of polyphenols, has been summarized. Following this, various members of different flavonoid classes have been individually discussed in the context of cancer. Members of flavanols, flavones, chalcones, flavanones and anthocyanins were

TABLE 5.7
Anthocyanins and the Summary of Various Proteins Influenced By Them in Neoplastic Cells

Anthocyanins	Target cancer	Protein targets mitigated	Targets escalated
Peonidin-3-glucoside	Colon cancer	HER2 phosphorylation	
	Lung cancer	u-PA, MMP2, MMP-9, p-ERK 1/2	
Black raspberry anthocyanins	Colon	DNMT3b, DNMT31, p-STAT3	SFRP2
Delphinidin	NSCLC	EGFR/VEGFR2, PI3K	
Black rice anthocyanins	Breast	FAK signalling, u-PA	
Cyanidin-3-O-glucoside	Glioma	MGMT and β-catenin	miR-214-5p

discussed thoroughly. The modulation of epigenetic mechanisms by these flavonoids for inducing anticancer effect has also been taken into consideration. It was noticed that these flavonoids inhibit or activate diverse molecular targets for provoking pharmacological effect. In a later chapter, the remaining types of plant-based HDACi, such as organosulfur compounds, quinones, alkaloids, prenylated isoflavones and other inhibitor types, will be focused on.

This segment on the top-down approach explained the different subgroups of flavonoids along with suitable representatives. Importantly, the anticancer activity of these flavonoid groups as monotherapeutic agents was highly stressed. The results indicate that, despite the ability of these flavonoids as single agents to induce cytotoxicity in cancer cells, their efficacy can be augmented by coupling these agents with standard anticancer chemotherapeutics. However, before discussing the combinatorial therapy with these inhibitors it is wise to emphasize first on non-flavonoid HDACi from natural sources and their single agent role as anticancer therapeutics.

REFERENCES

Afroze, N., Pramodh, S., Hussain, A., Waleed, M. & Vakharia, K. (2020). A review on myricetin as a potential therapeutic candidate for cancer prevention. *3 Biotech 10*: 211–211.

Almaguer, G., Ortiz-Vilchis, P., Cordero, P., Martinez-Vega, R., Perez-Durán, J., Meaney, E., Villarreal, F., Ceballos, G. & Nájera, N. (2021). Anticancer potential of (-)-epicatechin in a triple-negative mammary gland model. *J Pharm Pharmacol 73*: 1675–1682.

Almatroodi, S. A., Almatroudi, A., Khan, A. A., Alhumaydhi, F. A., Alsahli, M. A. & Rahmani, A. H. (2020). Potential therapeutic targets of epigallocatechin gallate (EGCG), the most abundant catechin in green tea, and its role in the therapy of various types of cancer. *Molecules 25*: 3146.

Alshatwi, A. A. (2010). Catechin hydrate suppresses MCF-7 proliferation through TP53/Caspase-mediated apoptosis. *Journal of Experimental & Clinical Cancer Research 29*: 167.

Alvarez, M. C., Maso, V., Torello, C. O., Ferro, K. P. & Saad, S. T. O. (2018). The polyphenol quercetin induces cell death in leukemia by targeting epigenetic regulators of pro-apoptotic genes. *Clinical epigenetics 10*: 139.

Angst, E., Park, J. L., Moro, A., Lu, Q.-Y., Lu, X., Li, G., King, J., Chen, M., Reber, H. A., Go, V. L. W., Eibl, G. & Hines, O. J. (2013). The flavonoid quercetin inhibits pancreatic cancer growth in vitro and in vivo. *Pancreas 42*: 223–229.

Antunes-Ricardo, M., Moreno-García, B. E., Gutiérrez-Uribe, J. A., Aráiz-Hernández, D., Alvarez, M. M. & Serna-Saldivar, S. O. (2014). Induction of apoptosis in colon cancer cells treated with isorhamnetin glycosides from Opuntia ficus-indica pads. *Plant Foods for Human Nutrition 69*: 331–336.

Arita, M., Koike, J., Yoshikawa, N., Kondo, M. & Hemmi, H. (2020). Licochalcone A inhibits BDNF and TrkB gene expression and hypoxic growth of human tumor cell lines. *International Journal of Molecular Sciences 21*: 506.

Badroon, N. A., Abdul Majid, N. & Alshawsh, M. A. (2020). Antiproliferative and apoptotic effects of cardamonin against hepatocellular carcinoma HepG2 cells. *Nutrients 12*: 1757.

Banjerdpongchai, R., Wudtiwai, B., Khaw-On, P., Rachakhom, W., Duangnil, N. & Kongtawelert, P. (2016). Hesperidin from citrus seed induces human hepatocellular carcinoma HepG2 cell apoptosis via both mitochondrial and death receptor pathways. *Tumour Biol 37*: 227–237.

Brown, O. I., Allgar, V. & Wong, K. Y. (2016). Coffee reduces the risk of death after acute myocardial infarction: a meta-analysis. *Coron Artery Dis 27*: 566–572.

Calogero, G., Sinopoli, A., Citro, I., Di Marco, G., Petrov, V., Diniz, A., Parola, A. J. & Pina, F. (2013). Synthetic analogues of anthocyanins as sensitizers for dye-sensitized solar cells. *Photochemical & Photobiological Sciences: Official Journal of the European Photochemistry Association and the European Society for Photobiology 12*: 883–894.

Chan, K. K. L., Siu, M. K. Y., Jiang, Y.-x., Wang, J.-j., Leung, T. H. Y. & Ngan, H. Y. S. (2018). Estrogen receptor modulators genistein, daidzein and ERB-041 inhibit cell migration, invasion, proliferation and sphere formation via modulation of FAK and PI3K/AKT signaling in ovarian cancer. *Cancer Cell Int 18*: 65.

Chang, H. L., Chang, Y. M., Lai, S. C., Chen, K. M., Wang, K. C., Chiu, T. T., Chang, F. H. & Hsu, L. S. (2017). Naringenin inhibits migration of lung cancer cells via the inhibition of matrix metalloproteinases-2 and -9. *Experimental and Therapeutic Medicine 13*: 739–744.

Chen, G., Zhou, D., Li, X. Z., Jiang, Z., Tan, C., Wei, X. Y., Ling, J., Jing, J., Liu, F. & Li, N. (2017). A natural chalcone induces apoptosis in lung cancer cells: 3D-QSAR, docking and an in vivo/vitro assay. *Sci Rep 7*: 10729.

Chen, H., Miao, Q., Geng, M., Liu, J., Hu, Y., Tian, L., Pan, J. & Yang, Y. (2013). Anti-tumor effect of rutin on human neuroblastoma cell lines through inducing G2/M Cell cycle arrest and promoting apoptosis. *The Scientific World Journal 2013*: 269165.

Chen, L., Jiang, B., Zhong, C., Guo, J., Zhang, L., Mu, T., Zhang, Q. & Bi, X. (2018). Chemoprevention of colorectal cancer by black raspberry anthocyanins involved the modulation of gut microbiota and SFRP2 demethylation. *Carcinogenesis 39*: 471–481.

Chen, M., Chen, Z., Huang, D., Sun, C., Xie, J., Chen, T., Zhao, X., Huang, Y., Li, D., Wu, B. & Wu, D. (2020). Myricetin inhibits TNF-α-induced inflammation in A549 cells via the SIRT1/NF-κB pathway. *Pulm Pharmacol Ther 65*: 102000.

Chen, X., Chen, S. & Yu, D. (2020). Protein kinase function of pyruvate kinase M2 and cancer. *Cancer Cell Int 20*: 523.

Chen, Y.-H., Wu, J.-X., Yang, S.-F., Yang, C.-K., Chen, T.-H. & Hsiao, Y.-H. (2022). Anticancer effects and molecular mechanisms of apigenin in cervical cancer cells. *Cancers (Basel) 14*: 1824.

Chevrollier, A., Loiseau, D., Reynier, P. & Stepien, G. (2011). Adenine nucleotide translocase 2 is a key mitochondrial protein in cancer metabolism. *Biochim Biophys Acta 1807*: 562–567.

Chi, Y. S., Lim, H., Park, H. & Kim, H. P. (2003). Effects of wogonin, a plant flavone from Scutellaria radix, on skin inflammation: in vivo regulation of inflammation-associated gene expression. *Biochem Pharmacol 66*: 1271–1278.

Chuang, C. Y., Tang, C. M., Ho, H. Y., Hsin, C. H., Weng, C. J., Yang, S. F., Chen, P. N. & Lin, C. W. (2019). Licochalcone A induces apoptotic cell death via JNK/p38 activation in human nasopharyngeal carcinoma cells. *Environ Toxicol 34*: 853–860.

Ciesielski, O., Biesiekierska, M. & Balcerczyk, A. (2020). Epigallocatechin-3-gallate (EGCG) alters histone acetylation and methylation and impacts chromatin architecture profile in human endothelial cells. *Molecules 25*: 2326.

Constantinescu, T. & Lungu, C. N. (2021). Anticancer activity of natural and synthetic chalcones. *International Journal of Molecular Sciences 22*: 11306.

de Pascual-Teresa, S., Santos-Buelga, C. & Rivas-Gonzalo, J. C. (2000). Quantitative analysis of flavan-3-ols in Spanish foodstuffs and beverages. *J Agric Food Chem 48*: 5331–5337.

Den Hartogh, D. J. & Tsiani, E. (2019). Antidiabetic properties of naringenin: a citrus fruit polyphenol. *Biomolecules 9*: 99.

Desmawati, D. & Sulastri, D. (2019). Phytoestrogens and their health effect. *Open Access Macedonian Journal of Medical Sciences 7*: 495–499.

Evans, L. W. & Ferguson, B. S. (2018). Food bioactive HDAC inhibitors in the epigenetic regulation of heart failure. *Nutrients 10*: 1120.

Farooqui, A. A. & Farooqui, T. (2018). Chapter 27 – Importance of fruit and vegetable-derived flavonoids in the Mediterranean diet: molecular and pathological aspects. *Role of the Mediterranean Diet in the Brain and Neurodegenerative Diseases*. T. Farooqui and A. A. Farooqui. Oxford: Academic Press: 417–427.

Fatima, N., Baqri, S. S. R., Bhattacharya, A., Koney, N. K.-K., Husain, K., Abbas, A. & Ansari, R. A. (2021). Role of flavonoids as epigenetic modulators in cancer prevention and therapy. *Frontiers in Genetics 12*: 758733.

Ganai, S. A. (2017). Plant-derived flavone apigenin: the small-molecule with promising activity against therapeutically resistant prostate cancer. *Biomed Pharmacother 85*: 47–56.

Ganai, S. A., Sheikh, F. A. & Baba, Z. A. (2021). Plant flavone chrysin as an emerging histone deacetylase inhibitor for prosperous epigenetic-based anticancer therapy. *Phytotherapy Research 35*: 823–834.

Ganai, S. A., Sheikh, F. A., Baba, Z. A., Mir, M. A., Mantoo, M. A. & Yatoo, M. A. (2021). Anticancer activity of the plant flavonoid luteolin against preclinical models of various cancers and insights on different signalling mechanisms modulated. *Phytother Res 35*: 3509–3532.

Gao, F., Li, M., Yu, X., Liu, W., Zhou, L. & Li, W. (2021). Licochalcone A inhibits EGFR signalling and translationally suppresses survivin expression in human cancer cells. *J Cell Mol Med 25*: 813–826.

Gao, W., Zhang, N. & Yu, S. (2014). Research progress on antitumor effects and mechanisms of phellinus. *Zhongguo Zhong yao za zhi= Zhongguo Zhongyao Zazhi= China Journal of Chinese Materia Medica 39*: 4165–4168.

Gao, Y., Yin, J., Rankin, G. O. & Chen, Y. C. (2018). Kaempferol induces G2/M cell cycle arrest via checkpoint kinase 2 and promotes apoptosis via death receptors in human ovarian carcinoma A2780/CP70 cells. *Molecules 23*: 1095.

Garg, A. & Chaturvedi, S. (2022). A comprehensive review on chrysin: emphasis on molecular targets, pharmacological actions and bio-pharmaceutical aspects. *Curr Drug Targets 23*: 420–436.

Geng, A., Xu, S., Yao, Y., Qian, Z., Wang, X., Sun, J., Zhang, J., Shi, F., Chen, Z., Zhang, W., Mao, Z., Lu, W. & Jiang, Y. (2022). Chrysin impairs genomic stability by suppressing DNA double-strand break repair in breast cancer cells. *Cell cycle (Georgetown, Tex.)* 21: 379–391.

Gong, G., Guan, Y.-Y., Zhang, Z.-L., Rahman, K., Wang, S.-J., Zhou, S., Luan, X. & Zhang, H. (2020). Isorhamnetin: a review of pharmacological effects. *Biomedicine & Pharmacotherapy 128*: 110301.

Graf, B. A., Milbury, P. E. & Blumberg, J. B. (2005). Flavonols, flavones, flavanones, and human health: epidemiological evidence. *Journal of Medicinal Food 8*: 281–290.

Gullón, B., Lú-Chau, T., Moreira, M., Lema, J. & Eibes, G. (2017). Rutin: a review on extraction, identification and purification methods, biological activities and approaches to enhance its bioavailability. *Trends in Food Science & Technology 67*: 220–235.

Guo, H., Ren, F., Zhang, L., Zhang, X., Yang, R., Xie, B., Li, Z., Hu, Z., Duan, Z. & Zhang, J. (2016). Kaempferol induces apoptosis in HepG2 cells via activation of the endoplasmic reticulum stress pathway. *Molecular Medicine Reports 13*: 2791–2800.

Guo, Y., Tong, Y., Zhu, H., Xiao, Y., Guo, H., Shang, L., Zheng, W., Ma, S., Liu, X. & Bai, Y. (2021). Quercetin suppresses pancreatic ductal adenocarcinoma progression via inhibition of SHH and TGF-β/Smad signaling pathways. *Cell Biol Toxicol 37*: 479–496.

Gutiérrez del Río Menéndez, I., Villar, C. J. & Lombó, F. (2016). Therapeutic uses of kaempferol: Anticancer and antiinflammatory activity. Biosynthesis, food sources and therapeutic uses *15.2*: 71–100.

Haghiac, M. & Walle, T. (2005). Quercetin induces necrosis and apoptosis in SCC-9 oral cancer cells. *Nutr Cancer 53*: 220–231.

Ham, S., Kim, K., Kwon, T., Bak, Y., Lee, D. H., Song, Y.-S., Park, S.-H., Park, Y., Kim, M., Kang, J., Hong, J. & Yoon, D.-Y. (2014). Luteolin induces intrinsic apoptosis via inhibition of E6/E7 oncogenes and activation of extrinsic and intrinsic signaling pathways in HPV-18-associated cells. *Oncol Rep 31*: 2683–2691.

Han, S., Lin, F., Qi, Y., Liu, C., Zhou, L., Xia, Y., Chen, K., Xing, J., Liu, Z., Yu, W., Zhang, Y., Zhou, X., Rao, T. & Cheng, F. (2022). HO-1 Contributes to luteolin-triggered ferroptosis in clear cell renal cell carcinoma via increasing the labile iron pool and promoting lipid peroxidation. *Oxid Med Cell Longev*: 3846217.

Hashemzaei, M., Delarami Far, A., Yari, A., Heravi, R. E., Tabrizian, K., Taghdisi, S. M., Sadegh, S. E., Tsarouhas, K., Kouretas, D., Tzanakakis, G., Nikitovic, D., Anisimov, N. Y., Spandidos, D. A., Tsatsakis, A. M. & Rezaee, R. (2017). Anticancer and apoptosis-inducing effects of quercetin in vitro and in vivo. *Oncol Rep 38*: 819–828.

Havsteen, B. H. (2002). The biochemistry and medical significance of the flavonoids. *Pharmacol Ther 96*: 67–202.

Ho, M. L., Chen, P. N., Chu, S. C., Kuo, D. Y., Kuo, W. H., Chen, J. Y. & Hsieh, Y. S. (2010). Peonidin 3-glucoside inhibits lung cancer metastasis by downregulation of proteinases activities and MAPK pathway. *Nutr Cancer 62*: 505–516.

Hong, S. H., Cha, H. J., Hwang-Bo, H., Kim, M. Y., Kim, S. Y., Ji, S. Y., Cheong, J., Park, C., Lee, H., Kim, G. Y., Moon, S. K., Yun, S. J., Chang, Y. C., Kim, W. J. & Choi, Y. H. (2019). Anti-proliferative and pro-apoptotic effects of licochalcone a through ROS-mediated cell cycle arrest and apoptosis in human bladder cancer cells. *International Journal of Molecular Sciences 20*: 3820.

Hostetler, G. L., Ralston, R. A. & Schwartz, S. J. (2017). Flavones: food sources, bioavailability, metabolism, and bioactivity. *Advances in Nutrition 8*: 423–435.

Hu, J., Zhang, Y., Jiang, X., Zhang, H., Gao, Z., Li, Y., Fu, R., Li, L., Li, J. & Cui, H. (2019). ROS-mediated activation and mitochondrial translocation of CaMKII contributes to Drp1-dependent mitochondrial fission and apoptosis in triple-negative breast cancer cells by isorhamnetin and chloroquine. *Journal of Experimental & Clinical Cancer Research 38*: 1–16.

Hu, S., Huang, L., Meng, L., Sun, H., Zhang, W. & Xu, Y. (2015). Isorhamnetin inhibits cell proliferation and induces apoptosis in breast cancer via Akt and mitogen-activated protein kinase kinase signaling pathways. *Molecular Medicine Reports 12*: 6745–6751.

Hua, F., Li, C. H., Chen, X. G. & Liu, X. P. (2018). Daidzein exerts anticancer activity towards SKOV3 human ovarian cancer cells by inducing apoptosis and cell cycle arrest, and inhibiting the Raf/MEK/ERK cascade. *Int J Mol Med 41*: 3485–3492.

Huang, L., Jin, K. & Lan, H. (2019). Luteolin inhibits cell cycle progression and induces apoptosis of breast cancer cells through downregulation of human telomerase reverse transcriptase. *Oncology letters 17*: 218.

Huang, W. C., Su, H. H., Fang, L. W., Wu, S. J. & Liou, C. J. (2019). Licochalcone A inhibits cellular motility by suppressing e-cadherin and MAPK signaling in breast cancer. *Cells 8*: 220–235.

Hussain, T., Gupta, S., Adhami, V. M. & Mukhtar, H. (2005). Green tea constituent epigallocatechin-3-gallate selectively inhibits COX-2 without affecting COX-1 expression in human prostate carcinoma cells. *Int J Cancer 113*: 660–669.

Imran, M., Rauf, A., Abu-Izneid, T., Nadeem, M., Shariati, M. A., Khan, I. A., Imran, A., Orhan, I. E., Rizwan, M., Atif, M., Gondal, T. A. & Mubarak, M. S. (2019). Luteolin, a flavonoid, as an anticancer agent: a review. *Biomed Pharmacother 112*: 108612.

Imran, M., Saeed, F., Gilani, S. A., Shariati, M. A., Imran, A., Afzaal, M., Atif, M., Tufail, T. & Anjum, F. M. (2021). Fisetin: an anticancer perspective. *Food Sci Nutr 9*: 3–16.

Ji, X., Liu, K., Li, Q., Shen, Q., Han, F., Ye, Q. & Zheng, C. (2022). A mini-review of flavone isomers apigenin and genistein in prostate cancer treatment. *Frontiers in Pharmacology 13*: 851589.

Jiang, C., Xiang, Y. & Zhong, Y. (2012). Effects of isorhamnetin on the proliferous cycle and apoptosis of human hepatoma HepG-2 cells: an experimental study. *J. Milit. Surg. Southwest China 14*: 432–435.

Jin, S., Zhang, Q. Y., Kang, X. M., Wang, J. X. & Zhao, W. H. (2010). Daidzein induces MCF-7 breast cancer cell apoptosis via the mitochondrial pathway. *Ann Oncol 21*: 263–268.

Jing, N., Song, J., Liu, Z., Wang, L. & Jiang, G. (2020). Glycosylation of anthocyanins enhances the apoptosis of colon cancer cells by handicapping energy metabolism. *BMC Complementary Medicine and Therapies 20*: 312.

Jung, Y. D., Kim, M. S., Shin, B. A., Chay, K. O., Ahn, B. W., Liu, W., Bucana, C. D., Gallick, G. E. & Ellis, L. M. (2001). EGCG, a major component of green tea, inhibits tumour growth by inhibiting VEGF induction in human colon carcinoma cells. *Br J Cancer 84*: 844–850.

Kashafi, E., Moradzadeh, M., Mohamadkhani, A. & Erfanian, S. (2017). Kaempferol increases apoptosis in human cervical cancer HeLa cells via PI3K/AKT and telomerase pathways. *Biomed Pharmacother 89*: 573–577.

Kaur, A., Sharma, Y., Kumar, A., Ghosh, M. P. & Bala, K. (2022). In-vitro antiproliferative efficacy of abrus precatorius seed extracts on cervical carcinoma. *Sci Rep 12*: 10226.

Khan, M. K., Zill, E. H. & Dangles, O. (2014). A comprehensive review on flavanones, the major citrus polyphenols. *Journal of Food Composition and Analysis 33*: 85–104.

Khoo, H. E., Azlan, A., Tang, S. T. & Lim, S. M. (2017). Anthocyanidins and anthocyanins: colored pigments as food, pharmaceutical ingredients, and the potential health benefits. *Food & Nutrition Research 61*: 1361779–1361779.

Kim, S. H., Hwang, K. A. & Choi, K. C. (2016). Treatment with kaempferol suppresses breast cancer cell growth caused by estrogen and triclosan in cellular and xenograft breast cancer models. *J Nutr Biochem 28*: 70–82.

Kim, S. O. & Kim, M. R. (2013). (-)-Epigallocatechin 3-gallate inhibits invasion by inducing the expression of Raf kinase inhibitor protein in AsPC-1 human pancreatic adenocarcinoma cells through the modulation of histone deacetylase activity. *Int J Oncol* 42: 349–358.

Kumar, V. & Chauhan, S. S. (2021). Daidzein induces intrinsic pathway of apoptosis along with ER α/β ratio alteration and ROS production. *Asian Pac J Cancer Prev* 22: 603–610.

Lai, S. L., Mustafa, M. R. & Wong, P. F. (2018). Panduratin A induces protective autophagy in melanoma via the AMPK and mTOR pathway. *Phytomedicine* 42: 144–151.

Lee, E. J., Oh, S. Y. & Sung, M. K. (2012). Luteolin exerts anti-tumor activity through the suppression of epidermal growth factor receptor-mediated pathway in MDA-MB-231 ER-negative breast cancer cells. *Food Chem Toxicol* 50: 4136–4143.

Lee, W.-J., Chen, Y.-R. & Tseng, T.-H. (2011). Quercetin induces FASL-related apoptosis, in part, through promotion of histone H3 acetylation in human leukemia HL-60 cells. *Oncol Rep* 25: 583–591.

Lee, W. J., Shim, J. Y. & Zhu, B. T. (2005). Mechanisms for the inhibition of DNA methyltransferases by tea catechins and bioflavonoids. *Mol Pharmacol* 68: 1018–1030.

Li, C., Yang, X., Hu, J. & Liao, J. (2012). Isorhamnetin suppresses the growth of gefitinib resistant human lung cancer PC9 cells. *Her. Med* 31: 831–834.

Li, H., Yang, B., Huang, J., Xiang, T., Yin, X., Wan, J., Luo, F., Zhang, L., Li, H. & Ren, G. (2013). Naringin inhibits growth potential of human triple-negative breast cancer cells by targeting β-catenin signaling pathway. *Toxicology letters* 220: 219–228.

Li, S., Yan, T., Deng, R., Jiang, X., Xiong, H., Wang, Y., Yu, Q., Wang, X., Chen, C. & Zhu, Y. (2017). Low dose of kaempferol suppresses the migration and invasion of triple-negative breast cancer cells by downregulating the activities of RhoA and Rac1. *OncoTargets and Therapy* 10: 4809–4819.

Li, T., Fu, X., Liu, B., Wang, X., Li, J., Zhu, P., Niu, X., Bai, J., Liu, Y., Lu, X. & Yu, Z. L. (2022). Luteolin binds Src, promotes STAT3 protein ubiquitination and exerts anti-melanoma effects in cell and mouse models. *Biochem Pharmacol* 200: 115044.

Li, X., Xu, J., Tang, X., Liu, Y., Yu, X., Wang, Z. & Liu, W. (2016). Anthocyanins inhibit trastuzumab-resistant breast cancer in vitro and in vivo. *Molecular Medicine Reports* 13: 4007–4013.

Li, Y., Wang, P. & Zhang, H. (2008). The inhibitory effect of isorhamnetin on growth of human gastric carcinoma cells. *Chin. Prim. Health Care* 6: 58–59.

Liang, Y. S., Qi, W. T., Guo, W., Wang, C. L., Hu, Z. B. & Li, A. K. (2018). Genistein and daidzein induce apoptosis of colon cancer cells by inhibiting the accumulation of lipid droplets. *Food & Nutrition Research* 62.

Liao, A. C., Kuo, C. C., Huang, Y. C., Yeh, C. W., Hseu, Y. C., Liu, J. Y. & Hsu, L. S. (2014). Naringenin inhibits migration of bladder cancer cells through downregulation of AKT and MMP-2. *Molecular Medicine Reports* 10: 1531–1536.

Lin, C.-H., Chang, C.-Y., Lee, K.-R., Lin, H.-J., Chen, T.-H. & Wan, L. (2015). Flavones inhibit breast cancer proliferation through the Akt/FOXO3a signaling pathway. *BMC Cancer* 15: 958–958.

Liu, D.-y., Guo, Y., Si, J.-y., Sun, G.-b., Zhang, B. & Cao, L. (2016). Inhibition of ovalitenin A on proliferation of HeLa cells via apoptosis, G2/M cell cycle arrest, and downregulation of COX-2. *Chinese Herbal Medicines* 8: 259–266.

Liu, Q., Cao, Y., Zhou, P., Gui, S., Wu, X., Xia, Y. & Tu, J. (2018). Panduratin A inhibits cell proliferation by inducing G0/G1 phase cell cycle arrest and induces apoptosis in breast cancer cells. *Biomol Ther (Seoul)* 26: 328–334.

Long, J., Li, B., Rodriguez-Blanco, J., Pastori, C., Volmar, C. H., Wahlestedt, C., Capobianco, A., Bai, F., Pei, X. H., Ayad, N. G. & Robbins, D. J. (2014). The BET bromodomain inhibitor I-BET151 acts downstream of smoothened protein to abrogate the growth of hedgehog protein-driven cancers. *J Biol Chem 289*: 35494–35502.

Luo, L.-P., Han, B., Yu, X.-P., Chen, X.-Y., Zhou, J., Chen, W., Zhu, Y.-F., Peng, X.-L., Zou, Q. & Li, S.-Y. (2014). Anti-metastasis activity of black rice anthocyanins against breast cancer: analyses using an ErbB2 positive breast cancer cell line and tumoral xenograft model. *Asian Pac J Cancer Prev 15*: 6219–6225.

Ma, Y. S., Yao, C. N., Liu, H. C., Yu, F. S., Lin, J. J., Lu, K. W., Liao, C. L., Chueh, F. S. & Chung, J. G. (2018). Quercetin induced apoptosis of human oral cancer SAS cells through mitochondria and endoplasmic reticulum mediated signaling pathways. *Oncology letters 15*: 9663–9672.

Määttä-Riihinen, K. R., Kähkönen, M. P., Törrönen, A. R. & Heinonen, I. M. (2005). Catechins and procyanidins in berries of vaccinium species and their antioxidant activity. *J Agric Food Chem 53*: 8485–8491.

Mahmood, N., Mihalcioiu, C. & Rabbani, S. A. (2018). Multifaceted role of the urokinase-type plasminogen activator (uPA) and its receptor (uPAR): diagnostic, prognostic, and therapeutic applications. *Front Oncol 8*: 24.

Manzolli, E. S., Serpeloni, J. M., Grotto, D., Bastos, J. K., Antunes, L. M. G., Barbosa Junior, F. & Barcelos, G. R. M. (2015). Protective effects of the flavonoid chrysin against methylmercury-induced genotoxicity and alterations of antioxidant status, in vivo. *Oxid Med Cell Longev*: 602360–602360.

Martens, S. & Mithöfer, A. (2005). Flavones and flavone synthases. *Phytochemistry 66*: 2399–2407.

Mazza, G. (1995). Anthocyanins in grapes and grape products. *Crit Rev Food Sci Nutr 35*: 341–371.

McCluskey, A. & Russell, C. (2021). Chalcones: potential anticancer agents. Translational Research in Cancer. DOI: 10.5772/intechopen.91441

Mehdi, S., Nafees, S., Zafaryab, M., Khan, M. & Rizvi, M. (2018). Chrysin: a promising anticancer agent its current trends and future perspectives. *European Journal of Experimental Biology 8*: 1–7.

Mirazimi, S. M. A., Dashti, F., Tobeiha, M., Shahini, A., Jafari, R., Khoddami, M., Sheida, A. H., EsnaAshari, P., Aflatoonian, A. H., Elikaii, F., Zakeri, M. S., Hamblin, M. R., Aghajani, M., Bavarsadkarimi, M. & Mirzaei, H. (2022). Application of quercetin in the treatment of gastrointestinal cancers. *Frontiers in Pharmacology 13*: 860209.

Nguyen, N. T., Nguyen, M. T., Nguyen, H. X., Dang, P. H., Dibwe, D. F., Esumi, H. & Awale, S. (2017). Constituents of the rhizomes of boesenbergia pandurata and their antiausterity activities against the PANC-1 human pancreatic cancer line. *Journal of Natural Products 80*: 141–148.

Oishi, M., Iizumi, Y., Taniguchi, T., Goi, W., Miki, T. & Sakai, T. (2013). Apigenin sensitizes prostate cancer cells to Apo2L/TRAIL by targeting adenine nucleotide translocase-2. *PLoS One 8*: e55922.

Oršolić, N., Štajcar, D. & Bašić, I. (2009). Propolis and its flavonoid compounds cause cytotoxicity on human urinary bladder transitional cell carcinoma in primary culture. *Periodicum biologorum 111*: 113–121.

Pal, H. C., Sharma, S., Strickland, L. R., Agarwal, J., Athar, M., Elmets, C. A. & Afaq, F. (2013). Delphinidin reduces cell proliferation and induces apoptosis of non-small-cell lung cancer cells by targeting EGFR/VEGFR2 signaling pathways. *PLoS One 8*: e77270.

Pal, S. & Saha, C. (2013). A review on structure–affinity relationship of dietary flavonoids with serum albumins. *J Biomol Struct Dyn 32*: 1132–1147.

Pal-Bhadra, M., Ramaiah, M. J., Reddy, T. L., Krishnan, A., Pushpavalli, S., Babu, K. S., Tiwari, A. K., Rao, J. M., Yadav, J. S. & Bhadra, U. (2012). Plant HDAC inhibitor chrysin arrest cell growth and induce p21WAF1by altering chromatin of STAT response element in A375 cells. *BMC Cancer 12*: 180.

Pan, J., Cai, X., Zheng, X., Zhu, X., Feng, J. & Wang, X. (2022). Luteolin inhibits viability, migration, angiogenesis and invasion of non-small cell lung cancer vascular endothelial cells via miR-133a-3p/purine rich element binding protein B-mediated MAPK and PI3K/Akt signaling pathways. *Tissue Cell 75*: 101740.

Panche, A. N., Diwan, A. D. & Chandra, S. R. (2016). Flavonoids: an overview. *Journal of Nutritional Science 5*: e47-e47.

Pandey, M., Kaur, P., Shukla, S., Abbas, A., Fu, P. & Gupta, S. (2012). Plant flavone apigenin inhibits HDAC and remodels chromatin to induce growth arrest and apoptosis in human prostate cancer cells: in vitro and in vivo study. *Mol Carcinog 51*: 952–962.

Paramanantham, A., Kim, M. J., Jung, E. J., Kim, H. J., Chang, S. H., Jung, J. M., Hong, S. C., Shin, S. C., Kim, G. S. & Lee, W. S. (2020). Anthocyanins isolated from vitis coignetiae pulliat enhances cisplatin sensitivity in MCF-7 human breast cancer cells through inhibition of Akt and NF-κB activation. *Molecules 25*: 3623.

Park, C., Cha, H.-J., Choi, E. O., Lee, H., Hwang-Bo, H., Ji, S. Y., Kim, M. Y., Kim, S. Y., Hong, S. H., Cheong, J., Kim, G.-Y., Yun, S. J., Hwang, H. J., Kim, W.-J. & Choi, Y. H. (2019). Isorhamnetin induces cell cycle arrest and apoptosis via reactive oxygen species-mediated AMP-activated protein kinase signaling pathway activation in human bladder cancer cells. *Cancers (Basel) 11*: 1494.

Park, H., Ra, J., Han, M. & Chung, J.-H. (2007). Hesperidin induces apoptosis in SNU-668, human gastric cancer cells. *Molecular and Cellular Toxicology 3*: 31–35.

Pascoal, A. C., Ehrenfried, C. A., Lopez, B. G., de Araujo, T. M., Pascoal, V. D., Gilioli, R., Anhê, G. F., Ruiz, A. L., Carvalho, J. E., Stefanello, M. E. & Salvador, M. J. (2014). Antiproliferative activity and induction of apoptosis in PC-3 cells by the chalcone cardamonin from Campomanesia adamantium (Myrtaceae) in a bioactivity-guided study. *Molecules 19*: 1843–1855.

Patel, K., Jain, A. & Patel, D. K. (2013). Medicinal significance, pharmacological activities, and analytical aspects of anthocyanidins 'delphinidin': a concise report. *Journal of Acute Disease 2*: 169–178.

Pramodh, S., Raina, R., Hussain, A., Bagabir, S. A., Haque, S., Raza, S. T., Ajmal, M. R., Behl, S. & Bhagavatula, D. (2022). Luteolin causes 5'CpG demethylation of the promoters of TSGs and modulates the aberrant histone modifications, restoring the expression of TSGs in human cancer cells. *International journal of molecular sciences 23*: 4067.

Predes, D., Oliveira, L. F. S., Ferreira, L. S. S., Maia, L. A., Delou, J. M. A., Faletti, A., Oliveira, I., Amado, N. G., Reis, A. H., Fraga, C. A. M., Kuster, R., Mendes, F. A., Borges, H. L. & Abreu, J. G. (2019). The chalcone lonchocarpin inhibits Wnt/β-Catenin signaling and suppresses colorectal cancer proliferation. *Cancers (Basel) 11*: 1968.

Prior, R. L. & Wu, X. (2006). Anthocyanins: structural characteristics that result in unique metabolic patterns and biological activities. *Free Radic Res 40*: 1014–1028.

Pu, Y., Zhang, T., Wang, J., Mao, Z., Duan, B., Long, Y., Xue, F., Liu, D., Liu, S. & Gao, Z. (2018). Luteolin exerts an anticancer effect on gastric cancer cells through multiple signaling pathways and regulating miRNAs. *Journal of Cancer 9*: 3669–3675.

Qin, Y., Cui, W., Yang, X. & Tong, B. (2016). Kaempferol inhibits the growth and metastasis of cholangiocarcinoma in vitro and in vivo. *Acta Biochimica et Biophysica Sinica 48*: 238–245.

Qiu, C., Zhang, T., Zhang, W., Zhou, L., Yu, B., Wang, W., Yang, Z., Liu, Z., Zou, P. & Liang, G. (2017). Licochalcone A inhibits the proliferation of human lung cancer cell lines A549 and H460 by inducing G2/M cell cycle arrest and ER stress. *International Journal of Molecular Sciences 18*: 1761.

Raina, R., Almutary, A. G., Bagabir, S. A., Afroze, N., Fagoonee, S., Haque, S. & Hussain, A. (2022). Chrysin modulates aberrant epigenetic variations and hampers migratory behavior of human cervical (HeLa) cells. *Frontiers in Genetics 12*: 768130–768130.

Raina, R., Pramodh, S., Rais, N., Haque, S., Shafarin, J., Bajbouj, K., Hamad, M. & Hussain, A. (2021). Luteolin inhibits proliferation, triggers apoptosis and modulates Akt/mTOR and MAP kinase pathways in HeLa cells. *Oncology Letters 21*: 192.

Rajoriya, S. (2021). 45P antimetastatic activity of apigenin in human breast carcinoma cells by regulation of different MMPs gene expression. *Annals of Oncology 32*: S18.

Ranganathan, S., Halagowder, D. & Sivasithambaram, N. D. (2015). Quercetin suppresses Twist to induce apoptosis in MCF-7 breast cancer cells. *PLoS One 10*: e0141370.

Rani, N., Bharti, S., Krishnamurthy, B., Bhatia, J., Sharma, C., Kamal, M. A., Ojha, S. & Arya, D. S. (2016). Pharmacological properties and therapeutic potential of naringenin: a citrus flavonoid of pharmaceutical promise. *Curr Pharm Des 22*: 4341–4359.

Ren, J., Lu, Y., Qian, Y., Chen, B., Wu, T. & Ji, G. (2019). Recent progress regarding kaempferol for the treatment of various diseases. *Experimental and Therapeutic Medicine 18*: 2759–2776.

Ren, Y., Wu, L., Li, X., Li, W., Yang, Y. & Zhang, M. (2015). FBXL10 contributes to the progression of nasopharyngeal carcinoma via involving in PI3K/mTOR pathway. *Neoplasma 62*: 925–931.

Rengarajan, T. & Yaacob, N. S. (2016). The flavonoid fisetin as an anticancer agent targeting the growth signaling pathways. *Eur J Pharmacol 789*: 8–16.

Robbins, R. & Bean, S. (2004). Development of a quantitative high-performance liquid chromatography-photodiode array detection measurement system for phenolic acids. *J Chromatogr A 1038*: 97–105.

Rong, W., Wan, N., Zheng, X., Shi, G., Jiang, C., Pan, K., Gao, M., Yin, Z., Gao, Z. J. & Zhang, J. (2022). Chrysin inhibits hepatocellular carcinoma progression through suppressing programmed death ligand 1 expression. *Phytomedicine 95*: 153867.

Rudrapal, M., Khan, J., Dukhyil, A. A. B., Alarousy, R. M. I. I., Attah, E. I., Sharma, T., Khairnar, S. J. & Bendale, A. R. (2021). Chalcone scaffolds, bioprecursors of flavonoids: chemistry, bioactivities, and pharmacokinetics. *Molecules 26*: 7177.

Sahu, N. K., Balbhadra, S. S., Choudhary, J. & Kohli, D. V. (2012). Exploring pharmacological significance of chalcone scaffold: a review. *Curr Med Chem 19*: 209–225.

Saleh, A., ElFayoumi, H., Youns, M. & Barakat, W. (2019). Rutin and orlistat produce antitumor effects via antioxidant and apoptotic actions. *Naunyn-Schmiedeberg's Archives of Pharmacology 392*: 165–175.

Samarghandian, S., Azimi-Nezhad, M., Borji, A., Hasanzadeh, M., Jabbari, F., Farkhondeh, T. & Samini, M. (2016). Inhibitory and cytotoxic activities of chrysin on human breast adenocarcinoma cells by induction of apoptosis. *Pharmacogn Mag 12*: S436–s440.

Sanaei, M., Kavoosi, F., Roustazadeh, A. & Golestan, F. (2018). Effect of genistein in comparison with trichostatin a on reactivation of DNMTs genes in hepatocellular carcinoma. *J Clin Transl Hepatol 6*: 141–146.

Semwal, D. K., Semwal, R. B., Combrinck, S. & Viljoen, A. (2016). Myricetin: a dietary molecule with diverse biological activities. *Nutrients 8*: 90.

Shan, S., Shi, J., Yang, P., Jia, B., Wu, H., Zhang, X. & Li, Z. (2017). Apigenin restrains colon cancer cell proliferation via targeted blocking of pyruvate kinase M2-dependent glycolysis. *Journal of Agricultural and Food Chemistry 65*: 8136–8144.

Shi, Y., Lian, K. & Jia, J. (2022). Apigenin suppresses the Warburg effect and stem-like properties in SOSP-9607 cells by inactivating the PI3K/Akt/mTOR signaling pathway. *Evidence-Based Complementary and Alternative Medicine 2022*: 3983637.

Shin, E., Choi, K.-M., Yoo, H.-S., Lee, C.-K., Hwang, B. Y., & Lee, M. K. (2010). Inhibitory effects of coumarins from the stem barks of *Fraxinus rhynchophylla* on adipocyte differentiation in 3T3-L1 cells. *Biological & Pharmaceutical Bulletin*, 33: 1610–1614. https://doi.org10.1248/bpb.33.1610

Shukla, S., Fu, P. & Gupta, S. (2014). Apigenin induces apoptosis by targeting inhibitor of apoptosis proteins and Ku70-Bax interaction in prostate cancer. *Apoptosis 19*: 883–894.

Shukla, S., Kanwal, R., Shankar, E., Datt, M., Chance, M. R., Fu, P., MacLennan, G. T. & Gupta, S. (2015). Apigenin blocks IKKα activation and suppresses prostate cancer progression. *Oncotarget 6*: 31216–31232.

Son, I., Chung, I.-M., Lee, S., Yang, H.-D. & Moon, H.-I. (2007a). Pomiferin, histone deacetylase inhibitor isolated from the fruits of Maclura pomifera. *Bioorganic & Medicinal Chemistry Letters 17*: 4753–4755.

Song, S., Su, Z., Xu, H., Niu, M., Chen, X., Min, H., Zhang, B., Sun, G., Xie, S., Wang, H. & Gao, Q. (2017). Luteolin selectively kills STAT3 highly activated gastric cancer cells through enhancing the binding of STAT3 to SHP-1. *Cell Death & Disease 8*: e2612–e2612.

Song, Y., Yu, J., Li, L., Wang, L., Dong, L., Xi, G., Lu, Y. J. & Li, Z. (2022). Luteolin impacts deoxyribonucleic acid repair by modulating the mitogen-activated protein kinase pathway in colorectal cancer. *Bioengineered 13*: 10998–11011.

Sun, H., Yin, M., Hao, D. & Shen, Y. (2020). Anti-cancer activity of catechin against A549 lung carcinoma cells by induction of cyclin kinase inhibitor p21 and suppression of cyclin E1 and P–AKT. *Applied Sciences 10*: 2065.

Sundaram, M. K., Ansari, M. Z., Al Mutery, A., Ashraf, M., Nasab, R., Rai, S., Rais, N. & Hussain, A. (2018). Genistein induces alterations of epigenetic modulatory signatures in human cervical cancer cells. *Anticancer Agents Med Chem 18*: 412–421.

Sundaram, M. K., Unni, S., Somvanshi, P., Bhardwaj, T., Mandal, R. K., Hussain, A. & Haque, S. (2019). Genistein modulates signaling pathways and targets several epigenetic markers in HeLa cells. *Genes 10*: 955.

Szkudelska, K. & Nogowski, L. (2007). Genistein – a dietary compound inducing hormonal and metabolic changes. *J Steroid Biochem Mol Biol 105*: 37–45.

Takahashi, Y., Iwai, M., Kawai, T., Arakawa, A., Ito, T., Sakurai-Yageta, M., Ito, A., Goto, A., Saito, M., Kasumi, F. & Murakami, Y. (2012). Aberrant expression of tumor suppressors CADM1 and 4.1B in invasive lesions of primary breast cancer. *Breast Cancer 19*: 242–252.

Ujiki, M. B., Ding, X.-Z., Salabat, M. R., Bentrem, D. J., Golkar, L., Milam, B., Talamonti, M. S., Bell, R. H., Jr., Iwamura, T. & Adrian, T. E. (2006). Apigenin inhibits pancreatic cancer cell proliferation through G2/M cell cycle arrest. *Molecular Cancer 5*: 76–76.

Valavanidis, A. & Vlachogianni, T. (2013). Chapter 8 – Plant polyphenols: recent advances in epidemiological research and other studies on cancer prevention. *Studies in Natural Products Chemistry*, vol. 39. Atta-ur-Rahman (ed.). Amsterdam: Elsevier: 269–295.

Waldecker, M., Kautenburger, T., Daumann, H., Busch, C. & Schrenk, D. (2008). Inhibition of histone-deacetylase activity by short-chain fatty acids and some polyphenol metabolites formed in the colon. *J Nutr Biochem 19*: 587–593.

Wang, J.-L., Quan, Q., Ji, R., Guo, X.-Y., Zhang, J.-M., Li, X. & Liu, Y.-G. (2018a). Isorhamnetin suppresses PANC-1 pancreatic cancer cell proliferation through S phase arrest. *Biomedicine & Pharmacotherapy 108*: 925–933.

Wang, R., Wang, J., Dong, T., Shen, J., Gao, X. & Zhou, J. (2019). Naringenin has a chemoprotective effect in MDA-MB-231 breast cancer cells via inhibition of caspase-3 and -9 activities. *Oncology Letters 17*: 1217–1222.

Wang, R., Yang, L., Li, S., Ye, D., Yang, L., Liu, Q., Zhao, Z., Cai, Q., Tan, J. & Li, X. (2018). Quercetin inhibits breast cancer stem cells via downregulation of aldehyde dehydrogenase 1A1 (ALDH1A1), chemokine receptor type 4 (CXCR4), mucin 1 (MUC1), and epithelial cell adhesion molecule (EpCAM). *Medical Science Monitor: International Medical Journal of Experimental and Clinical Research 24*: 412–420.

Wang, S.-H., Wu, C.-H., Tsai, C.-C., Chen, T.-Y., Tsai, K.-J., Hung, C.-M., Hsu, C.-Y., Wu, C.-W. & Hsieh, T.-H. (2022). Effects of luteolin on human breast cancer using gene expression array: inferring novel genes. *Current Issues in Molecular Biology 44*: 2107–2121.

Wang, S.-M., Yang, P.-W., Feng, X.-J., Zhu, Y.-W., Qiu, F.-J., Hu, X.-D. & Zhang, S.-H. (2021). Apigenin inhibits the growth of hepatocellular carcinoma cells by affecting the expression of microRNA transcriptome. *Front Oncol 11*: 657665.

Wang, Z., Liu, H., Hu, Q., Shi, L., Lü, M., Deng, M. & Luo, G. (2021). Cardamonin inhibits the progression of oesophageal cancer by inhibiting the PI3K/AKT signalling pathway. *Journal of Cancer 12*: 3597–3610.

WenXing, L., YiHong, B. & ZhenYu, W. (2011). Morphological observation of apoptotic human colon cancer cells HT29 induced by anthocyanins from Lonicera edulis. *Acta Nutrimenta Sinica 33*: 575–579.

Wojdyło, A., Oszmiański, J. & Czemerys, R. (2007). Antioxidant activity and phenolic compounds in 32 selected herbs. *Food Chemistry 105*: 940–949.

Wong, N., De Melo, J. & Tang, D. (2013). PKM2, a central point of regulation in cancer metabolism. *International Journal of Cell Biology 2013*: 242513.

Wu, H., Huang, M., Liu, Y., Shu, Y. & Liu, P. (2015). Luteolin induces apoptosis by up-regulating miR-34a in human gastric cancer cells. *Technol Cancer Res Treat 14*: 747–755.

Wu, H., Pan, L., Gao, C., Xu, H., Li, Y., Zhang, L., Ma, L., Meng, L., Sun, X. & Qin, H. (2019). Quercetin inhibits the proliferation of glycolysis-addicted HCC cells by reducing hexokinase 2 and Akt-mTOR pathway. *Molecules 24*: 1993.

Wu, H.-T., Lin, J., Liu, Y.-E., Chen, H.-F., Hsu, K.-W., Lin, S.-H., Peng, K.-Y., Lin, K.-J., Hsieh, C.-C. & Chen, D.-R. (2021). Luteolin suppresses androgen receptor-positive triple-negative breast cancer cell proliferation and metastasis by epigenetic regulation of MMP9 expression via the AKT/mTOR signaling pathway. *Phytomedicine 81*: 153437.

Xu, Q., Gao, B., Liu, X., Zhang, X., Wu, L., Xing, D., Ma, L. & Liu, J. (2022). Myocyte enhancer factor 2D promotes hepatocellular carcinoma through AMOTL2/YAP signaling that inhibited by luteolin. *Int J Clin Exp Pathol 15*: 206–214.

Xu, Y., Tong, Y., Ying, J., Lei, Z., Wan, L., Zhu, X., Ye, F., Mao, P., Wu, X., Pan, R., Peng, B., Liu, Y. & Zhu, J. (2018). Chrysin induces cell growth arrest, apoptosis, and ER stress and inhibits the activation of STAT3 through the generation of ROS in bladder cancer cells. *Oncology Letters 15*: 9117–9125.

Yan, X., Hao, Y., Chen, S., Jia, G., Guo, Y., Zhang, G., Wang, C., Cheng, R., Hu, T., Zhang, X. & Ji, H. (2019). Rutin induces apoptosis via P53 up-regulation in human glioma CHME cells. *Translational Cancer Research 8*: 2005–2013.

Yang, C., Wang, Z., Tao, D. & Peng, T. (2003). Effect of isorhamnetin on bcl-2 gene expression of HeLa cell. *Med J West China 1*: 196–198.

Yao, S., Wang, X., Li, C., Zhao, T., Jin, H. & Fang, W. (2016). Kaempferol inhibits cell proliferation and glycolysis in esophagus squamous cell carcinoma via targeting EGFR signaling pathway. *Tumour Biol 37*: 10247–10256.

Yi, X., Zuo, J., Tan, C., Xian, S., Luo, C., Chen, S., Yu, L. & Luo, Y. (2016). Kaempferol, a flavonoid compound from gynura medica induced apoptosis and growth inhibition in MCF-7 breast cancer cell. *African Journal of Traditional, Complementary, and Alternative Medicines: AJTCAM 13*: 210–215.

Yuan, L. W., Jiang, X. M., Xu, Y. L., Huang, M. Y., Chen, Y. C., Yu, W. B., Su, M. X., Ye, Z. H., Chen, X., Wang, Y. & Lu, J. J. (2021). Licochalcone A inhibits interferon-gamma-induced programmed death-ligand 1 in lung cancer cells. *Phytomedicine 80*: 153394.

Yumnam, S., Park, H. S., Kim, M. K., Nagappan, A., Hong, G. E., Lee, H. J., Lee, W. S., Kim, E. H., Cho, J. H., Shin, S. C. & Kim, G. S. (2014). Hesperidin induces paraptosis like cell death in hepatoblastoma, HepG2 cells: involvement of ERK1/2 MAPK [corrected]. *PLoS One 9*: e101321–e101321.

Zahra, K., Dey, T., Ashish, Mishra, S. P. & Pandey, U. (2020). Pyruvate kinase M2 and cancer: the role of PKM2 in promoting tumorigenesis. *Front Oncol 10*: 159.

Zaidun, N. H., Thent, Z. C. & Latiff, A. A. (2018). Combating oxidative stress disorders with citrus flavonoid: naringenin. *Life Sci 208*: 111–122.

Zang, Y. Q., Feng, Y. Y., Luo, Y. H., Zhai, Y. Q., Ju, X. Y., Feng, Y. C., Wang, J. R., Yu, C. Q. & Jin, C. H. (2019). Glycitein induces reactive oxygen species-dependent apoptosis and G0/G1 cell cycle arrest through the MAPK/STAT3/NF-κB pathway in human gastric cancer cells. *Drug Dev Res 80*: 573–584.

Zeng, L., Zhen, Y., Chen, Y., Zou, L., Zhang, Y., Hu, F., Feng, J., Shen, J. & Wei, B. (2014). Naringin inhibits growth and induces apoptosis by a mechanism dependent on reduced activation of NF-κB/COX-2-caspase-1 pathway in HeLa cervical cancer cells. *Int J Oncol 45*: 1929–1936.

Zhai, T., Zhang, X., Hei, Z., Jin, L., Han, C., Ko, A. T., Yu, X. & Wang, J. (2021). Isorhamnetin inhibits human gallbladder cancer cell proliferation and metastasis via PI3K/AKT signaling pathway inactivation. *Frontiers in Pharmacology 12*: 628621.

Zhang, B., Su, J.-P., Bai, Y., Li, J. & Liu, Y.-H. (2015). Inhibitory effects of O-methylated isoflavone glycitein on human breast cancer SKBR-3 cells. *Int J Clin Exp Pathol 8*: 7809–7817.

Zhang, H., Zhong, X., Zhang, X., Shang, D., Zhou, Y. I. & Zhang, C. (2016). Enhanced anticancer effect of ABT-737 in combination with naringenin on gastric cancer cells. *Experimental and Therapeutic Medicine 11*: 669–673.

Zheng, W., Sun, R., Yang, L., Zeng, X., Xue, Y. & An, R. (2017). Daidzein inhibits chorio-carcinoma proliferation by arresting cell cycle at G1 phase through suppressing ERK pathway in vitro and in vivo. *Oncol Rep 38*: 2518–2524.

Zhou, J., Zhu, Y. F., Chen, X. Y., Han, B., Li, F., Chen, J. Y., Peng, X. L., Luo, L. P., Chen, W. & Yu, X. P. (2017). Black rice-derived anthocyanins inhibit HER-2-positive breast cancer epithelial-mesenchymal transition-mediated metastasis in vitro by suppressing FAK signaling. *Int J Mol Med 40*: 1649–1656.

Zhou, Y., Chen, L., Ding, D., Li, Z., Cheng, L., You, Q. & Zhang, S. (2022). Cyanidin-3-O-glucoside inhibits the β-catenin/MGMT pathway by upregulating miR-214-5p to reverse chemotherapy resistance in glioma cells. *Scientific Reports 12*: 7773.

Zhu, L. & Xue, L. (2019). Kaempferol suppresses proliferation and induces cell cycle arrest, apoptosis, and DNA damage in breast cancer cells. *Oncol Res 27*: 629–634.

6 Non-Flavonoid Histone Deacetylase Inhibitors from Natural Sources and Their Stupendous Anticancer Properties

Doubtlessly, the major proportion of plant-isolated histone deacetylase inhibitors (Hdaci) comes within the boundary of flavonoids. Certain non-flavonoid molecules derived from plants are emerging as phenomenal anticancer molecules with congenital histone deacetylase inhibitory activity. These non-flavonoid HDACi are either isothiocyanates, bromo-tyrosines, stilbenes, organosulfur compounds and coumarins. Sulforaphane, the HDAC inhibitor of the isothiocyanate group, is among the most well-studied natural molecules from the perspective of cancer. Bis (4-hydroxybenzyl) sulfide, diallyl disulfide (DADS), S-allyl mercaptocysteine (SAMC) and diallyl trisulfide (DATS) falling within the group of organosulfur compounds have also demonstrated HDAC blocking potential. Propionate, valeric acid and butyrate that belong to the short chain fatty acid group are also non-flavonoid molecules possessing the HDAC obstructing property. Piceatannol and resveratrol fall within the confines of stilbenes. While the bromo-tyrosine derivative group includes psammaplins, dihydrocoumarin comes under the banner of coumarins.

6.1 PROMISING ANTICANCER EFFECT OF ISOTHIOCYANATES

Sulforaphane and other isothiocyanates induce anticancer signalling through a variety of mechanisms. In the HCT116 cell line, sulforaphane inhibited various HDACs including HDAC3 and HDAC6 (Dickinson et al. 2015). This isothiocyanate prompted apoptosis in bladder cancer cells through the intrinsic apoptotic mechanism. Following treatment with sulforaphane a substantial increase in ROS levels occurred, which were soothed when cells were subjected to antioxidant treatment with N-acetyl-L-cysteine. Activation of the nuclear factor-E2-related factor-2 (Nrf2) and endoplasmic reticulum stress was confirmed post-sulforaphane treatment (Jo et al. 2014). Pharmacological advantages of sulforaphane were studied on hepatocellular carcinoma models. This molecule by way of down-regulating 6-phosphofructo-2-kinase/fructose-2, 6-biphosphatase4 (PFKFB4) elicited cytotoxicity in these cells (Jeon et al. 2011). Antileukaemic activity has been ascribed to sulforaphane. The induction of apoptosis was recorded in acute lymphoblastic leukaemia cells after proper treatment with the defined isothiocyanate. In addition to cell cycle arrest, the inhibition of AKT

DOI: 10.1201/9781003294863-6

and mTOR pathways was deciphered in cells mixed with sulforaphane. The tumour burden in all xenograft models was lowered following sulforaphane administration, suggesting that this promising anticancer dietary molecule works well both under *in vitro* and *in vivo* conditions (Suppipat et al. 2012).

Rhabdomyosarcoma (RMS), the soft-tissue sarcoma, has two subtypes –alveolar (ARMS) and embryonal (ERMS). Sulforaphane induced growth inhibition in both subtypes in a time- and dose-dependent manner. In ARMS, intervention was accompanied with down-regulation of MYCN, MET, PAX3-FKHR and down-regulation of DR5 and p21 (Bergantin et al. 2014). Sulforaphane inhibited invasion and showed strong induction of apoptosis in glioblastoma cells dose dependently. While the treatment lowered the levels of survivin and Bcl-2, an increase in expression of Bax, Bad and cytochrome C was confirmed. The invasion inhibitory effects of this isothiocyanate on glioblastoma cells were mediated by the decline in the levels of Galectin-3, MMP-2, MMP-9 and the elevation in E-cadherin (Zhang et al. 2016). Sulforaphane has demonstrated mind-blowing anticancer effect on cervical cancer cells. Cell cycle arrest, cyclin B1 down-regulation and proliferation inhibition were observed in cells exposed to sulforaphane (Cheng, Tsai & Hsu 2016). Glioblastoma cells develop drug resistance due to more survival signals and defective apoptotic pathways. Glioblastoma stem cells also have key importance in the recurrence and development of glioblastoma. Sulforaphane is considered to have a multitargeted effect and, as such, has been evaluated against glioblastoma and glioblastoma stem cells. Isothiocyanate (sulforaphane) induced apoptotic signalling and reduced survival of glioblastoma cells through caspase activation, ROS generation, double and single strand DNA breaks. Sulforaphane proved to be successful in removing glioblastoma stem cells having a central role in the recurrence and resistance of glioblastoma. The administration of this inhibitor showed promising results against ectopic xenografts. The growth of these xenografts was markedly reduced by 100 mg/kg of sulforaphane given for a three-week time period (Bijangi-Vishehsaraei et al. 2017).

The efficacy of sulforaphane has been evaluated against breast cancer stem cells. These were inhibited by sulforaphane and this phenomenon was accompanied with the inhibition of Wnt/beta-catenin pathway (Li et al. 2010). Breast cancer metastasis is considered the premier concern of cancer-related mortality in women. Isomers of sulforaphane S-sulforaphane and R-sulforaphane substantially attenuated migration and invasion of breast cancer cells induced by TGF-β1. Also sulforaphane inhibited phosphorylation of ERK and MEK and thus this outcome signifies that isothiocyanate sulforaphane obstructs RAF/MEK/ERK signalling (Zhang et al. 2022). Another study assessed the therapeutic effects of sulforaphane on multiple breast cancer lines with the aim to confirm whether it exerts HDAC inhibitory function in these cells. As usual, this plant-derived molecule not only restrained cell growth but also induced cell cycle arrest at G2-M and enhanced cyclin B1 expression. Fas ligand induction in MDA-MB-231 cells on sulforaphane intervention played a critical role in triggering apoptotic death. Fas induction in turn activated certain caspases, namely caspase-8 and 3 and PARP. On the other hand, in the remaining three breast cancer cell lines the down-regulated expression of Bcl-2 kick-started apoptosis. The activity of HDACs was inhibited by sulforaphane and other molecular targets, including estrogen receptor-A, human epidermal growth factor receptor-2, epidermal growth factor receptor were down-regulated in all cell lines but, despite HDAC inhibition, no change was seen in the acetylation status of histones H3 and H4 (Pledgie-Tracy, Sobolewski & Davidson 2007). Glioblastoma multiforme, the bellicose brain tumour, is not sensitive to the majority of therapeutic approaches (Osuka & Van Meir 2017). The profound necrosis and microvascular hyperplasia that occur in this tumour are elicited by hypoxia. The antineoplastic properties of sulforaphane have been studied under hypoxia and normoxia in glioblastoma cells. Evidence-based study suggests that this HDAC inhibitor has the strength to induce apoptosis in these cells both under conditions of hypoxia or normoxia. The activation of caspase 3 and 7, fragmentation of DNA was observed in treated cells. Sulforaphane blocked the cell cycle at the synthetic phase and thus inhibited the proliferation of the said cells. These therapeutic effects of sulforaphane have been partly ascribed to its ability to evoke oxidative stress by way of lowering levels of glutathione and by modulation of extracellular signal-regulated kinases through their increased phosphorylation (Sita et al. 2021).

Small molecules from plants have shown promising results against different maladies encompassing cancer. These natural compounds are gaining significance in cancer chemoprevention. Sulforaphane modifies epigenetic mechanisms in cells and thus exerts anticancer effects. Many cancers, including oral cancer, have epigenetic aetiology and thus sulforaphane, being a plant-derived molecule, has been examined for prevention of oral cancer. Three distinct concentrations of sulforaphane were inspected on oral squamous cell carcinoma cells for two different time points. Sulforaphane markedly inhibited HDAC activity and proliferation of these cells. The lower concentration induced cell cycle arrest while the two higher concentrations dragged cells towards apoptosis. Sulforaphane alleviated the levels of MMP in a time- and dose-related fashion and promoted apoptosis via intrinsic and extrinsic apoptotic routes (Krishnan et al. 2022). It is a fact that sulforaphane exerts antiinflammatory, neuroprotective and antineoplastic effects. While evaluating this inhibitor on oral cancer cells, it was delineated that sulforaphane hampered the invasiveness and motility of these cells (SCC-14 and SCC-9) by depleting the expression of cathepsin s. Further, LC3 conversion was elevated by sulforaphane and the siRNA knockdown of the former facilitated the tendency of cell migration (Chen et al. 2018).

Considered among the most deadly tumours, pancreatic ductal adenocarcinoma in its advanced stage is difficult to tackle as good therapeutic options are not available. The antiinflammatory molecule sulforaphane has proved effective against multiple tumour entities. While under typical conditions miR135b-5p and the gene which it regulates, namely RASAL2, were found to be expressed, their expression in cancer lines and malignant pancreatic tissues was nearly lacking. Pharmacological intervention with sulforaphane in BxPC-3 cells showed up-regulation of RASAL2 (tumour suppressor) through induction of miR135b-5p and thus attenuated pancreatic cancer progression (Yin et al. 2019). Moreover, it was observed that the introduction of miR135b-5p through lipofection also raised the expression of RASAL2, further confirming that the latter is positively regulated by the former. The therapeutic potential of sulforaphane was assessed in acute myeloid leukaemia patients and cell lines in the context of miR-181a (small non-coding RNA). In the first half of the study, the levels of miR-181a were quantified in acute myeloid leukaemia patients in comparison to controls (healthy individuals) in order to dissect the role of this miR in the pathogenesis of the aforementioned disease. In the second half, the effect of sulforaphane was unravelled on acute myeloid leukaemia lines, keeping growth, proliferation and apoptosis of these cells in focus. Compared to healthy controls, the expression of miR-181a in acute myeloid leukaemia patients demonstrated a 2.9 fold increase. The antiproliferative effect was confirmed in acute myeloid leukaemia cells post-exposure to sulforaphane. This antiproliferative effect was noticed in association with lowered expression of miR-181a (Koolivand et al. 2022). From these findings one can infer that the overexpression of miR-181a has a critical role in the aetiology of acute myeloid leukaemia and sulforaphane-induced proliferation inhibition of these cells is mediated through down-regulation of this miR. Necroptosis and ferroptosis are considered as non-apoptotic death pathways offering great therapeutic strength. The non-apoptotic cell death induced by sulforaphane has been explored, and it was revealed that this isothiocyanate dose-dependently prompts ferroptosis and apoptosis in acute myeloid leukaemia (AML) cells. While sulforaphane at elevated concentration induced ferroptosis that was associated with a decline in the GSH (intracellular) and GSH peroxidase 4 protein, caspase-reliant apoptosis was recorded when this inhibitor was applied to cells at lower concentration (Greco et al. 2021). As it is quite established that pro-apoptotic mechanisms are resisted by malignant cells, thus sulforaphane having the potential to provoke different types of cell deaths may prove an effective antitumour agent.

6.2 PRODIGIOUS CYTOTOXIC EFFECT OF BROMOTYROSINE HDAC INHIBITORS

Such type of HDACi described in the previous chapter are derived from sponges. These inhibitors have inherent antimicrobial and anticancer activity (Kim et al. 1999). Psammaplin A exerts inhibitory activity against HDACs and is most well studied among psammaplins with reference to cancer. This inhibitor has been examined on endometrial cancer cells for its therapeutic properties. Dose-dependent inhibition of Ishikawa cell proliferation was quantified once these cells were given treatment with sulforaphane. This treatment was accompanied with cyclin-dependent kinase inhibitor up-regulation and down-regulation of certain proteins namely CDKs, cyclins, pRb culminating in cell cycle arrest and subsequent apoptosis (Ahn et al. 2008). Atypical levels of HDACs create cancer onset and in many cases advancement. Among the various psammaplin derivatives, psammaplin A exhibited strong HDAC inhibitory activity. This inhibitor induced gelsolin at transcript level through hyperacetylation of histone proteins. Compared to non-reduced forms of psammaplin A, a reduced version showed relatively stronger HDAC inhibition (Kim, Shin & Kwon 2007).

Psammaplin A has been certified to inhibit certain HDACs and DNA methyltransferase. Studies on the distribution of this inhibitor have shown its main distribution in lungs, suggesting that the prodrug psammaplin A may prove beneficial in cancers related to lungs (Kim et al. 2012). Actually, psammaplin A is not a monomer but a bromotyrosine disulfide dimer, but it acts in the form of a reduced monomer (Psa-SH). This reduced monomer inhibits all the class I HDACs except HDAC8. Psa-SH treatment resulted in apoptosis of HL-60 cells by inducing histone

H3 hyperacetylation. Apart from the haematological cell line, the defined form of psammaplin A hampered growth of various solid cancer lines under *in vitro* conditions. Further, inhibition of growth was noticed under *in vivo* conditions in Lewis lung cancer cells (Bao et al. 2021). Despite the radiosensitizer property of psammaplin, its low bioavailability restrains its clinical uses. Thus, attempts have been made to generate new radiosensitizers using psammaplin A as the molecule of departure. Nine molecules were synthesized using psammaplin A as scaffold. Among them, eight demonstrated cytotoxic effect to a variable extent against glioblastoma and lung cancer cells. While six compounds were able to radiosensitize glioblastoma cells, only four were capable to do so in lung cancer cells. Two compounds were successful in radiosensitizing both lines and among them one compound demonstrated substantial inhibitory activity against DNMT1 (Wee et al. 2019).

Aminopeptidase N has implications in angiogenesis and invasion of tumour cells. Psammaplin A inhibits the activity of this peptidase through a non-competitive mechanism. This bromotyrosine derivative showed antiproliferative property against endothelial and multiple cancer lines. Tube formation as well as invasion of endothelial cells induced by fibroblast growth factor was curbed by psammaplin A (Shim et al. 2004). These results suggest that psammaplin A is a novel aminopeptidase N inhibitor that has the potential to shrink tumour growth through blockade of angiogenesis.

6.3 CYTOTOXIC EFFECTS OF ORGANOSULFUR COMPOUNDS

Certain organosulfur compounds encompassing bis (4-hydroxybenzyl) sulfide have manifested HDAC inhibitory character and are thus evaluated for possible anticancer effect. Separated from the root extract of a flowering plant (*Pleuropterus ciliinervis*), this compound showed strong HDAC inhibitory tendency. Bis (4-hydroxybenzyl) sulfide evinced growth inhibitory effect on several tumour cell lines, but its effect was more profound on the breast cancer (MDA-MB-231)-line (Son et al. 2007). However, the main impediment with organosulfur compounds is the concern about their stability. Diallyl disulfide (DADS) is another member of organosulfur group of histone deacetylase inhibitors. Mounting evidence supports the notion that garlic inhibits proliferation in a variety of cancer cells. This bulb contains both oil-soluble and water-soluble sulphur compounds. Water-soluble compounds, such as diallyl trisulfide (DATS), diallyl disulfide (DADS) and diallyl sulfide (DAS), offer more cancer protection than water-insoluble ones. The main organosulfur from garlic, namely DADS, inhibits the burgeoning of many cancer types, This compound facilitates enzymes involved in the detoxification of cancer-causing agents, inhibits DNA adduct formation, scavenges free radicals (Yi & Su 2013). The prophylactic role of DADS has been studied in rodent models in which skin carcinogenesis was induced through application of DMBA (cutaneously) and TPA. It was found that DADS hampered the incidence of skin tumour and multiplicity in mice. This effect was attributed to the enhancement of activities of various antioxidant enzymes and Nrf2 accumulation in the nucleus. Additionally, in mice with Nrf2 knockout skin carcinogenesis was developed as this could counteract the effect of DADS. Scrutinization at molecular level unbossomed that DADS facilitated endogenous binding between Nrf2 and p21 thereby impeding the Nrf2 degradation mediated by Keap1 (Shan et al. 2016).

DADS was examined on osteosarcoma (MG-63) cells at different concentrations and for three different time durations. The viability of these cells was reduced by DADS following a time- and dose-dependent trend. With the increasing concentration of DADS, the apoptotic ratio elevated and the cells were arrested at stage G2/M. It was authenticated that DADS induced autophagy in these cells as high levels of LC3-II were recorded in treated cells. Also, the apoptosis induced by DADS was impaired when an autophagy inhibitor 3-methyladenine was used. Further, the kinase activity of p-mTOR was suppressed by DADS resulting in inhibition of PI3K/Akt/mTOR signalling (Heckmann et al. 2013; Yue et al. 2019). Collectively, this outcome suggests that DADS induce cell cycle arrest in osteosarcoma cells and activate apoptotic and autophagic death in these cells through blockade of mTOR signalling. Ehrlich ascites carcinoma (EAC)-tumour model was employed to explore the antitumour potential of DADS. A substantial reduction in tumour growth was seen and it was also noticed that DADS facilitate expression of caspase-3, prevent antitumour molecules like p53 from oxidative degradation, up-regulate SOD and NQO1 (antioxidant enzymes) (Puttalingaiah et al. 2017).

Parkinsonism associated deglycase (DJ-1) is overexpressed in several cancers such as breast, liver, lung, oesophageal and pancreatic cancer (Zhang et al. 2008; Wu, Liang & Huang 2009; Wei et al. 2013). Its elevated expression facilitates oncogenesis and lowers the antiproliferative effect of cancer chemotherapeutics. Nuclear expression of DJ-1 prevents apoptosis, facilitates proliferation of HL-60 cells and increases their migration ability. Treatment with DADS considerably reduced the expression of DJ-1 at protein level in these cells. In addition to this, p-Fak and p-Src were substantially inhibited by DADS in HL-60 cells (Liu et al. 2018). Thus, the migration and invasion of HL-60 cells (DJ-1 high expression) is inhibited by DADS through attenuation of Src signalling. Squamous cell carcinoma of oesophagus is a bellicose tumour with high mortality across the globe. The viability of these cells was markedly reduced by DADS time and dose dependently. Cell cycle arrest and apoptosis was induced by DADS in the aforesaid cells. The cytotoxic effect of DADS was accompanied with the activation of caspases, an increase in Bax/Bcl-2 ratio and inhibition of MEK-ERK signalling (Yin et al. 2014).

Effective therapies are not available for prostate cancer, which is not androgen independent. DADS was investigated for its antiproliferative and cell cycle blocking ability against prostate cancer cells. Following a dose-dependent trend DADS impaired the prostate cancer cell growth. Cell cycle arrest was triggered in these cells by DADS by way of mitigating the expression of CDK1 (Arunkumar et al. 2006). DADS, the premier component of garlic, exert multiple biological effects. Colon carcinogenesis induced by chemicals is reversed by DADS. The impact of DADS on acetylation of histones was explored in HT-29 and Caco-2 cell lines. Further, the influence of allyl mercaptan, the metabolite of DADS, was examined in HDACs context. While reduction in HDAC activity with DADS (200 µM) was found to be 29%, the decline in activity at identical concentration of allyl mercaptan was estimated to be 92%. In these cell lines DADS elicited enhanced expression of p21 (waf1/cip1) at both transcript and at protein level (Druesne et al. 2004). On the whole, DADS exert antiproliferative property against defined cell lines through obstruction of HDAC activity, subsequent hyperacetylation of histones and by heightening the p21 (waf1/cip1) expression.

Among the various organosulfur compounds scrutinized, allyl mercaptan displayed the strongest HDAC inhibitory potential. This organosulfur compound demonstrated K_i value of 24 μM against HDAC8. Colon cancer cells subjected to exposure of this inhibitor induced acetylation of histones. As revealed by the chromatin immunoprecipitation technique, allyl mercaptan hyperacetylated histone H3 on the promoter of P21WAF1, resulting in enhanced Sp3 binding. This establishes that Sp3 plays a premier role in the expression of P21 following inhibition of HDACs by allyl mercaptan. Allyl mercaptan arrested cells at the G1 phase of cell cycle and this arrest corresponded to expression of p21Waf1 (Nian et al. 2008). Another active molecule isolated from garlic is S-allylmercaptocysteine (SAMC). Under *in vitro* condition this compound has portrayed antitumour effect. SAMC has been assessed on colorectal carcinoma (SW620) cells. The defined cysteine depleted cell viability by stimulating apoptotic death in these cells. SAMC modulated p38 and JNK signalling, enhanced p53 protein and activated Bax in SW620 (Zhang et al. 2014). This entire outcome suggests that SAMC exerts apoptotic and inhibits multiplication of colorectal cancer cells through alteration of JNK as well as p38 pathway.

SAMC and S-allylcysteine were tested on a pair of colon cancer lines (HT-29 and SW-480) for their cell cycle and antiproliferative effects. Among these garlic derivatives, SAMC proved to be effective in inhibiting the growth of these lines and provoked apoptosis. A marked elevation in the reduced glutathione levels were recorded in SAMC applied cells (Shirin et al. 2001). SAMC also evinced inhibitory effects on the growth of mouse erythroleukemia (DS19)- cells. This inhibitor also promoted histone acetylation in breast (T47D) and colon cancer (Caco-2) cells (Lea et al. 2002). A plethora of evidence suggests that the constituents of garlic exhibit anticancer properties. SAMC was checked for its apoptotic ability in anaplastic thyroid carcinoma (8305C) cells. Intervention with SAMC dose dependently resulted in induction of cytotoxicity (apoptosis) in predefined cells, changed their morphological features and restrained the activity of telomerase (Liu et al. 2015). Inhibitor of differentiation (Id-1) has severe implications in bladder cancer as its high expression has been observed in a variety of cancers. Apart from this, the higher expression of Id-1 makes the bladder cancer cells resistant to drugs and thus declines their sensitivity to therapeutic agents (Zhao et al. 2020). SAMC, the aqueous soluble constituent of garlic, has shown activity against multitude cancer types (Lv et al. 2019; Ferrari 2022). Experiments were done on bladder cancer cells to probe the link of SAMC-invoked apoptosis with expression of Id-1. SAMC-induced apoptosis was impeded by a high expression of Id-1 while its inactivation sensitized the cells to the above-mentioned garlic derivative. The suppression of invasion and also migration of bladder cancer cells due to SAMC occurred through Id-1 down-regulation (Hu et al. 2011). SAMC considered as the aqueous soluble compound isolated from garlic has thoroughly investigated using estrogen receptor negative (MDA-MB-231) and estrogen receptor positive (MCF-7) breast cancer lines. In both types of cells SAMC displayed growth inhibition through the induction of G0/G1 phase arrest. Blockade of the cell cycle was accompanied with p21 and p53 up-regulation. SAMC treatment culminated in apoptosis of these cells as evidenced by the nuclear morphology changes, Bax activation, down-regulation of Bcl-X$_L$, Bcl-2 ensuing caspase-9/3 activation (Zhang et al. 2014).

One more organosulfur compound isolated from garlic with antineoplastic property is diallyl trisulfide (DATS). This derivative controls various processes linked to cancer including angiogenesis, cell cycle, metastasis, apoptosis and invasion (Puccinelli & Stan 2017). While evaluating the DATS in DLD-1 and HCT-15 (colon cancer lines) substantial suppression of cell growth was noticed. The formation of a microtubule network in these cells was disrupted by this trisulfide (Seki et al. 2008). A reduction in viability of gastric cancer (BGC-823) cells was seen once these cells were given DATS treatment. Induction of cell cycle arrest that coincided with cyclin B1 and A2 accumulation was also peculiar in trisulfide-exposed cells. DATS stimulated apoptosis in the aforesaid cells through caspase activation and by modulating Bcl-2 family members. DATS administration in xenograft (BGC-823) collapsed tumour growth and significantly depleted the number of Ki-67 (proliferation marker) bearing cells in tumours. Further DATS under both conditions markedly activated p38, JNK/MAPK and hampered Nrf2/Akt signalling (Jiang et al. 2017). DATS substantially depleted the viability of osteosarcoma cells, lessened cyclin D1 and induced G0/G1 arrest. Moreover, DATS promoted ROS production and disrupted mitochondrial membrane potential in these MNNG/HOS and MG63 cells. This trisulfide facilitated apoptosis in these cells through obstruction of the PI3K/Akt pathway. The DATS-induced effects were reversed by N-acetylcysteine, which is a known ROS scavenger (Wang et al. 2016). Thus, apoptosis induced by DATS in osteosarcoma cells is ROS intervened and is accomplished through inhibition of PI3K/Akt signalling.

6.4 ANTICANCER CHARACTERISTICS OF NATURAL SHORT CHAIN FATTY ACIDS

Among this group of inhibitors butyrate is extensively studied. Butyrate is produced in the intestines when dietary fibre undergoes microbial fermentation. This fatty acid facilitates the apoptosis of cancer cells via inhibition of SIR3. Butyrate reverses the Warburg effect through induction of PDHA1 hyperacetylation (Xu et al. 2017). Butyrate provoked an apoptotic type of cell death in cancer cells and attenuated tumour advancement through inhibition of HDACs. In a couple of colon cancer lines butyrate was assessed for its pharmacological effect and it was observed that this inhibitor results in Caspase-1 cleavage in Caco-2 cells while in RSB cells expression of this caspase was not detectable. While in the former cell line butyrate induced enhanced expression of active subunit of caspase-3, cleavage of this caspase was noticed in the latter line. In both cell lines butyrate treatment was associated with increased expression of cleaved PARP. The expression of Bax underwent up-regulation while Bcl-2 was down-regulated following the exposure of Caco-2 cells to butyrate. Z-YVAD-FMK induced caspase-1 inactivation repealed butyrate invoked effects in these cells and this was not the case with RSB cells. In the latter cell line the effects of butyrate were abrogated through caspase-3 inactivation by Z-DVED-FMK (Avivi-Green et al. 2002). Butyrate demonstrated cytotoxic effect against colorectal cancer cells through activation of p38 MAPK and by stimulating caspase 3/7 (Fung et al. 2011). These findings suggest that butyrate follows similar but not completely identical mechanisms for inducing apoptosis in different types of colon cancer cells. Butyrate produced in the colon by beneficial microbes exhibits stunning anticancer activity. While butyrate is the premier energy source for typical colonocytes, it exerts cell growth inhibitory effects on cancerous colonocytes. In colorectal cancer cells, butyrate not only blocked glucose transport but also glycolysis. This occurred by via depletion of GLUT1 in membrane and glucose-6-phosphate dehydrogenase in cytoplasm (Geng et al. 2021). Cancer-accompanied stromatogenesis in a mouse model of pancreatic ductal adenocarcinoma was significantly alleviated by butyrate. This HDAC inhibitor also safeguarded intestinal mucosa integrity and lowered certain microorganisms known to be pro-inflammatory (Panebianco et al. 2022).

Another member of this group, namely valeric acid, has manifested significant activity against cancer cells. Proliferation of breast cancer cells was lowered by valeric acid intervention. The inhibitory effect was more profound in MCF-7 and MDA-MB-231 over the MCF-10A cell line. The colony formation capacity of MDA-MB-231 and MCF-7 cells was attenuated when valeric acid was applied to them as per standard procedure. Valeric acid inhibited HDAC activity in MCF-7 cells in concentration-dependent fashion. An inhibition of 40.44%, 53.15% and 61.64% was recorded at 2.5 mM, 5mM and 10 mM concentrations of valeric acid respectively. The global decline in DNA methylation was also quantified from treated cells as compared to the control (Shi et al. 2021). This outcome points towards the conclusion that valeric acid declines proliferation, migration and invasion of breast cancer cells through inhibition of HDACs and by lowering DNA methylation globally. In other words, valeric acid opts the epigenetic route for inducing antiproliferative effect in breast cancer cells. Valeric acid has been explored for its therapeutic benefit against liver cancer. The defined acid demonstrated particularly high cytotoxic effect against liver cancer cells, inhibited their proliferation and cell invasion. In short valeric acid hampers liver cancer development through its HDAC blocking ability (Han et al. 2020). Radiation exposure, either medical or accidental, induces intestinal and haematopoietic side-effects. Out of the short chain fatty acids, valeric acid demonstrated highly substantial radioprotection. Treatment with valeric acid not only enhanced irradiated mice survival but also rectified the function of their gastrointestinal tract, safeguarded their haematogenic organs. Irradiated mice exhibited down-regulation of Keratin, type II cytoskeletal 1 (KRT1) and this declined expression was abrogated on valeric acid administration (Li et al. 2020).

6.5   ANTIPROLIFERATIVE EFFECTS OF STILBENES

They are considered as non-flavonoid polyphenols. These polyphenols are peculiar in having 1, 2-diphenylethylene nucleus. Due to their marvellous antiinflammatory, antioxidant and abnormal cell death activation characteristics, stilbenes are emerging as amazing agents for treatment and prevention of various maladies including cancer (Sirerol et al. 2016). Actually, they are secondary metabolites of plants, which are synthesized through shikimate pathway. Resveratrol and piceatannol are two highly studied stilbenes and their pharmacokinetic and physiochemical concerns need to be subdued prior to their applicability (Navarro-Orcajada et al. 2022).

6.5.1 Resveratrol as Antiproliferative Stilbene

Resveratrol, the most thoroughly explored stilbene, occurs in grapes, peanuts, blue-berries, apples and plums (Koushki et al. 2018). While exploring the chemopreventive strength of this stilbene in bladder cancer cells, it was confirmed that it substantially reduces the viability of these cells. This effect occurred through caspase 9 and 3 activation besides the modulation of certain proteins of the Bcl-2 family. Resveratrol intervention was accompanied with cell cycle arrest at the G1 phase, down-regulation of p-Rb, p-Akt, cyclin-dependent kinase 4 and cyclin D1. On the other hand, resveratrol induced p21 activation and p38 MAPK phosphorylation. In the xenograft model of bladder cancer treatment, resveratrol lowered the expression of fibroblast growth factor-2 and vascular endothelial growth factor (Bai et al. 2010). Experimental evidence supports that resveratrol is highly effective against colo357 and capan-2 (pancreatic cancer lines). These cell lines manifested more sensitivity towards resveratrol treatment. Importantly, no substantial toxicity was exerted by resveratrol in normal pancreatic cells, thereby signifying its clinical relevance. The above-defined cancer lines evinced marked caspase-3 activation in addition to down-regulation of p21 and p53 (Zhou et al. 2011).

Anchorage independent and dependent growth of COLO 201 and HT-29 (colon cancer) cells were obstructed by resveratrol. This growth inhibition was associated with caspase 3 and caspase 8 cleavage. In the latter cell line, resveratrol induced autophagy that was characterized through autophagic vacuoles and increased LC3-II. Autophagy induced by this stilbene when blocked by 3-methyladenine (autophagy inhibitor) markedly declined apoptosis by inhibiting caspase 8 and 3 cleavage suggesting the cytotoxic effect of resveratrol stimulated autophagy. Further, obstructing resveratrol-invoked apoptosis through Z-VAD (OMe)-FMK (pan-caspase inhibitor) failed to decline autophagy but raised LC3-II. Increased ROS generation induced by resveratrol corresponded with caspase8/3 cleavage and escalated LC3-II levels (Miki et al. 2012). This effect of resveratrol was attenuated on exposing cells to N-acetylcysteine clearly indicating that resveratrol-promoted apoptosis by way of autophagy in colon cancer cells is ROS interceded. Resveratrol was tested on a pair of follicular thyroid carcinoma cell lines and a couple of papillary thyroid carcinoma lines. The treated cells showed shuttling of MAPK to nucleus, up-regulated p53, its phosphorylation and transcript levels of p21, c-jun and c-fos. These resveratrol-induced effects were counteracted on blockade of the MAPK pathway through PD 98059 or antisense transfection of H-ras antisense transfection (Shih et al. 2002).

A study was designed in which it was hypothesized whether resveratrol restrains proliferation of multiple myeloma cells through modulation of STAT3 and NF-κB pathways. The proliferation of multiple myeloma cells was suppressed by resveratrol irrespective of their sensitive or resistant behaviour towards standard chemotherapeutic agents. Resveratrol invoked apoptotic death in multiple myeloma cells through inhibition of the NF-κB pathway, which is constitutively active. Additionally resveratrol soothed induced (interleukin-6 triggered) and constitutive STAT3 activation (Bhardwaj et al. 2006). Pharmacological intervention involving resveratrol dragged breast cancer cells to apoptosis. This was accompanied with the release of cytochrome c from mitochondria to cytosol. The redox state of these cells was also modified by resveratrol and it was observed that this stilbene triggered early ROS generation and oxidation of lipids (Filomeni et al. 2007). Also, resveratrol was confirmed to commit oral squamous cell carcinoma cells to apoptosis. Following its treatment, the viability of SCC25, SCC15 and CAL-27 cells reduced in a dose and time-dependent manner. The apoptotic mechanism followed the mitochondrial pathway and the inhibitory effect of resveratrol on migration and the invasion of oral squamous OSCC cells has been attributed to its ability to inhibit transcription factors favouring epithelial to mesenchymal transition (Kim et al. 2018).

Using ovarian cancer cells as models, the molecular mechanism modulated by resveratrol in these cells for inducing anticancer effect was unbossomed. The importance of autophagy in resveratrol-induced apoptosis in these cells was also taken into account. It was delineated that resveratrol impelled ROS production in ovarian cancer cells induces autophagy, which in turn is succeeded by apoptosis. Stilbene resveratrol facilitated cleavage of LC3 and up-regulated the expression of ATG5. Genetic and pharmacological blockade of autophagy pacified the resveratrol-stimulated apoptotic death. Chloroquine, the inhibitor of autophagy involved late in the process of autophagy, markedly reduced death and activation of caspase-3 induced by this non-flavonoid polyphenol. The knockdown of ATG5 through siRNA also bridled apoptotic cell death promoted by resveratrol (Lang et al. 2015). Conclusively for provoking apoptosis in ovarian cancer lines resveratrol requires the assistance of ROS production and autophagy. The multiple effects of resveratrol exposure were recorded in T-cell acute lymphoblastic leukemia (T-ALL) cells. Besides, the inhibition of cell proliferation, apoptotic death and autophagy was induced by resveratrol in these cells. Through down-regulation of cyclin D1 and cyclin A and up-regulation of p27 and p21, resveratrol elicited cell cycle arrest. Besides, resveratrol declined the levels of Bcl-2 and Mcl-1, the antiapoptotic proteins and concurrently up-regulated the Bad, Bim and Bax, which are well-established proapoptotic proteins. Treated cells showed considerable enhancement in LC3-II/LC3-I and in the expression of Beclin 1. Resveratrol triggered considerable Akt, 4E-BP1, p70S6K and mTOR dephosphorylation but increased p38-MAPK phosphorylation. Resveratrol-elicited apoptosis was heightened when the above-mentioned cells were treated with a proven autophagy inhibitor, namely 3-methyladenine (Ge et al. 2013). The conclusion that can be derived from the above results is that resveratrol modulates molecular players of multiple pathways for inducing apoptosis in T-ALL cells. Resveratrol decreased colon cancer cell viability by up-regulating apoptotic, autophagic and endoplasmic reticulum stress markers. The impact of resveratrol on apoptosis of these cells was partly abolished on the use of 3-methyladenine (Jia et al. 2019).

It has been proved that resveratrol induces apoptosis in pancreatic and breast cancer models through the modulation of Nrf2. Resveratrol induces translocation of this transcription factor to the nucleus where it enhances antioxidant (free radical scavenger) gene expression (Alavi et al. 2021). Apart from its cancer-preventive properties, resveratrol also possesses cancer-quelling properties. It has been experimentally verified that resveratrol provokes senescence in lung and breast cancer cells. This effect has been credited to the tendency of resveratrol to stimulate ER-stress, facilitating expression p38-MAPK, declining nitric oxide level to elevate the expression of deleted in liver cancer 1 (DLC1). Further, ER-stress induced DNA damage and dysfunction of mitochondria resulting in senescence of cancer cells (Bian et al. 2022). In various cancer lines including two breast cancer lines cellular senescence induced by this stilbene has been connected to DLC1. This study has concluded that oxidative stress facilitated by resveratrol plays a key role in directing DLC1-intervened senescence in atypical (cancer) cells (Ji et al. 2018). Furthermore, proliferation and viability of hepatocellular carcinoma cells was alleviated by keeping these cells in a resveratrol environment. Resveratrol not only enhanced expression of SIRT1 at the protein stage but also increased its enzymatic function. This resulted in the eventual decline in p-PI3K, p-AKT and p-Forkhead Box O3a in these cells. The blockade of SIR1 activity through EX527 reversed the effects of the former on the above-mentioned kinases (Chai et al. 2017). This confirms that SIR1 activation elicited by resveratrol is critical for curbing the PI3K/AKT pathway. Certain microRNAs have implications in cancer. Only recently it has been found that over-representation of miR-155-5p facilitates genesis of cancer, including cervical (Li et al. 2019). Overexpression of this miR was quantified from gastric cancer cells and clinical samples. Resveratrol-based intervention lowered the expression profile of miR-155-5p and thereby inhibited growth of gastric cancer cells. The hyperexpression of miR-155-5p made the conditions conducive for proliferation as well as invasion and metastasis of these cells. Also this abnormal expression blocked apoptosis of gastric cancer cells (Su et al. 2022). These results signify that resveratrol induced therapeutic effects in gastric cancer cells are executed through alleviated expression of miR-155-5p.

6.5.2 Piceatannol in Abrogating Cancer Advancement

Multiple preclinical studies confirmed the strength of piceatannol to attenuate the growth of a variety of cancers of distinct origins (Banik et al. 2020). The growth of a pair of melanoma cell lines was stopped by piceatannol application. It was further confirmed that cells treated with this polyphenol evinced markedly raised expression of miR-181a and this miR has Bcl-2 as its target gene. The effects of piceatannol were reversed on miR-181a silencing. This output confirms that piceatannol impairs the growth of melanoma cells by way of heightening the expression of miR-181a (Du, Zhang & Gao 2017). The proliferation and invasion induced by VEGF in human umbilical vein endothelial cells was significantly attenuated by piceatannol. In embryos of zebrafish, this stilbene considerable blocked the subintestinal vessel formation. VEGF-induced ROS generation was also soothed by this non-flavonoid polyphenol. In addition, proliferation as well as migration induced by VEGF in colon cancer cells was also subdued by piceatannol. The angiogenic functions favoured by VEGF are attenuated by piceatannol (Hu et al. 2020). Another study has also confirmed the promising effect of piceatannol in thwarting the progression of colon cancer (Chiou et al. 2022). The pharmacological advantage of piceatannol has been studied on azoxymethane/dextran sulfate sodium-elicited growth of colon cancer cells. The administration of piceatannol reduced the tumour area by 57.2%, tumour numbers by 30.1% and the number of Ki-67 positive cells by 89.1%. In addition to this, PD-1 and colon MCP-1 levels were depleted by 70.9% and 43.8% respectively. Last, but no way least, piceatannol reduced the number of COX-2 positive cells up to 60.2%. Taken together, these results indicate that colon tumour growth inhibition induced by piceatannol occurs via down-regulation of levels of colon MCP-1 and PD-1 and also has cross-talk with COX-2 expression (Kimura 2022).

6.6 COUMARINS IN ANTICANCER THERAPY

Coumarins predominantly occur in medicinal plants and they come under the banner of the benzopyrone family. Natural coumarins exert broad spectrum therapeutic activities, such as anticancer, antiinflammatory, antibacterial, antifungal, antimalarial, neuroprotective, antiviral, antihypertensive and anticonvulsant (Sharifi-Rad et al. 2021). Coumarins are classified as simple coumarins, including ostruthin, novobiocin, ammoresinol, esculetin, coumarin, esculin, phellodenol A, umbelliferone, fraxidin and fraxin; furano coumarins encompassing psoralen, imperatorin, methoxsalen, bergapten; dihydrofurano coumarins holding felamidin, anthogenol and marmesin under its umbrella; bicoumarins covering dicumarol; linear type including xanthyletin, agasyllin and aegelinol benzoate; angular coumarins engirdling pseudocordatolide C, (+)-dihydrocalanolide A and B and some more (Shin et al. 2010; Witaicenis, Seito & Di Stasi 2010; Sharifi-Rad et al. 2021).

Radiotherapeutic and chemotherapeutic agents cause DNA damage in cancer cells, which is repaired by effectual repair mechanisms operating in these cells. Targetting DNA repair mechanisms in cancer cells have emerged as a propitious tactic to improve the anticancer potential of various anticancer agents. Dihydrocoumarin, a member of coumarins, impairs the repair of DNA double-strand breaks in unicellular fungi. This coumarin inhibits Rad52 after DNA damage, which creates double strand break deficiencies. Deletion of gene encoding Class I HDAC namely *RPD3* reproduced the effect of dihydrocoumarin on Rad52, suggesting that the effects of this coumarin are mediated through inhibition of defined HDAC (Chen et al. 2017). These findings suggest that dihydrocoumarin may augment the anticancer potential of other therapeutic agents by obstructing DNA repair and committing the cells towards apoptosis.

Osthole isolated from *Cnidium monnieri Cusson* (fruit of this plant) is a natural derivative of coumarin. Emerging findings suggest the anticancer activity of this coumarin derivative. In breast cancer cells, osthole has been found to be effectual in obstructing migration as well as invasion. This derivative obstructed not only the promoter activity of metalloproteinase-2 but also its enzymatic function. This may be among the reasons resulting in blockade of the above-mentioned processes by osthole (Yang et al. 2010). The bottom line is that osthole targets MMP-2 in breast cancer cells for obstructing their migration and invasion. Osthole has also been tested on human medulloblastoma (TE671) and laryngeal cancer (RK33) lines. The dose-dependent decline in proliferation and viability was estimated in these cells following the intervention with osthole. Moreover, osthole exerted apoptotic death in this cells that was associated increased cell number in G1 phase (Jarząb et al. 2014). The pharmacological activity of osthole was evaluated on rat glioma cells. Osthole inhibited the proliferation of these cells and induced apoptosis through up-regulation and down-regulation of proapoptotic proteins and antiapoptotic proteins respectively. Most importantly, this natural derivative productively impaired migration and invasion of these cells in comparison to corresponding control. Certain signalling pathways, namely MAPK and PI3K/Akt, were inhibited by osthole (Ding et al. 2013). These results show that osthole may prove effective in circumventing glioma growth and progression.

Osthole was checked for its therapeutic effects on hypoxic HCT116 (colon cancer) cells and the underlying mechanism followed by this anticancer molecule was explored. The migration and viability of these cells were markedly lowered by osthole. The activation of phospho-eukaryotic initiation factor 2 alpha and elevation of pro-apoptotic players was also recorded in osthole-treated cells (Peng & Chou 2022). An interesting finding was obtained when osthole was assessed on cervical cancer (HeLa) cells. Osthole induced both apoptosis and pyroptosis in these cells. In pyroptosis the integrity of cell membrane is lost and the release of lactate dehydrogenase occurs. The critical protein in both these processes is caspase-3. Z-DEVD-FMK, the irreversible inhibitor of caspase-3 rescued the osthole-induced effects (apoptosis and pyroptosis) in these cells. Gasdermin E-intervened pyroptosis and apoptosis and it belongs to the gasdermin family of proteins. This protein is stimulated by therapeutic agents to induce pyroptosis in abnormal (cancer) cells and invoke immunity against tumour (antitumour immunity) (Liao et al. 2022). Its knockdown in HeLa cells only alleviated pyroptosis but demonstrated no effect on apoptosis. NAD(P)H quinone oxidoreductase 1 (NQO1) is well known for carrying out reduction (two electron) of quinones (Pey, Megarity & Timson 2019). Cervical cancer cells on intervention with osthole manifested reduction in NQO1 expression, subsequent increase in ROS generation culminating in apoptotic death and pyroptosis. In the xenograft model of cervical cancer, osthole shrinked tumour growth and reduced NQO1 expression (Wang et al. 2022). Speaking concisely, osthole elicited pharmacological effects in cervical cancer models through caspase-3, gasdermin E activation and NQO1 down-regulation-arbitrated amassing of ROS. Esculetin, a simple coumarin, blocked the proliferation and invasion of laryngeal cancer cells and depleted tumour weight and tumour growth in the xenograft model dose dependently. Apart from inducing cell cycle arrest at the G1/S phase, this coumarin considerably decreased phosphorylation of STAT3 and obstructed its shuttling to the cell nucleus (Zhang, Xu & Zhou 2019). Thus, esculetin exhibits antiproliferative, antimigratory and antiinvasive effect in laryngeal cancer cell through induction of cell cycle arrest and by modulating the phosphorylation status of STAT3 and its subcellular localization.

This part of the book has thoroughly summarized non-flavonoid HDACi from natural origin in the context of cancer. A compendium of broccoli-isolated sulforaphane and its promising activity against various cancer types has been taken into consideration. Following this, the bromotyrosine derivative, namely psammaplin A, was discussed along with its antineoplastic effects and the underlying molecular-signalling mechanisms modified. Interestingly, multiple organosulfur compounds encompassing bis (4-hydroxybenzyl) sulfide, DADS, DATS and SAMC were extensively considered and their pharmacological effects and their molecular targets in multiple cancer types were unbossomed. Afterwards, natural short chain fatty acid members, including butyrate and valeric acid, were debated with respect to various cancers and the cellular players activated or inhibited by them were also taken into consideration. Then stilbenes such as resveratrol and piceatannol and their therapeutic benefit in the inhibition of cancer cell proliferation, migration as well as invasion, tumour growth along with the signalling mechanisms modulated by them in distinct cancer types, were intensely explained. Last, but no way least, various coumarin types along with their corresponding members were classified, following which the counter cancer activities of certain coumarins, especially osthole, were clearly elaborated in the context of the most recent and mounting experimental corroborations. Thus, the forthcoming chapter will illuminate the importance and heightened therapeutic benefits of plant-derived flavonoid HDACi when used in conjunction with conventional therapeutic agents.

REFERENCES

Ahn, M. Y., Jung, J. H., Na, Y. J. & Kim, H. S. (2008). A natural histone deacetylase inhibitor, psammaplin A, induces cell cycle arrest and apoptosis in human endometrial cancer cells. *Gynecol Oncol 108*: 27–33.

Alavi, M., Farkhondeh, T., Aschner, M. & Samarghandian, S. (2021). Resveratrol mediates its anti-cancer effects by Nrf2 signaling pathway activation. *Cancer Cell Int 21*: 579.

Arunkumar, A., Vijayababu, M. R., Srinivasan, N., Aruldhas, M. M. & Arunakaran, J. (2006). Garlic compound, diallyl disulfide induces cell cycle arrest in prostate cancer cell line PC-3. *Mol Cell Biochem 288*: 107–113.

Avivi-Green, C., Polak-Charcon, S., Madar, Z. & Schwartz, B. (2002). Different molecular events account for butyrate-induced apoptosis in two human colon cancer cell lines. *J Nutr 132*: 1812–1818.

Bai, Y., Mao, Q. Q., Qin, J., Zheng, X. Y., Wang, Y. B., Yang, K., Shen, H. F. & Xie, L. P. (2010). Resveratrol induces apoptosis and cell cycle arrest of human T24 bladder cancer cells in vitro and inhibits tumor growth in vivo. *Cancer Sci 101*: 488–493.

Banik, K., Ranaware, A. M., Harsha, C., Nitesh, T., Girisa, S., Deshpande, V., Fan, L., Nalawade, S. P., Sethi, G. & Kunnumakkara, A. B. (2020). Piceatannol: a natural stilbene for the prevention and treatment of cancer. *Pharmacol Res 153*: 104635.

Bao, Y., Xu, Q., Wang, L., Wei, Y., Hu, B., Wang, J., Liu, D., Zhao, L. & Jing, Y. (2021). Studying histone deacetylase inhibition and apoptosis induction of psammaplin A monomers with modified thiol group. *ACS Med Chem Lett 12*: 39–47.

Bergantin, E., Quarta, C., Nanni, C., Fanti, S., Pession, A., Cantelli-Forti, G., Tonelli, R. & Hrelia, P. (2014). Sulforaphane induces apoptosis in rhabdomyosarcoma and restores TRAIL-sensitivity in the aggressive alveolar subtype leading to tumor elimination in mice. *Cancer Biology & Therapy 15*: 1219–1225.

Bhardwaj, A., Sethi, G., Vadhan-Raj, Saroj, Bueso-Ramos, C., Takada, Y., Gaur, U., Nair, A. S., Shishodia, S. & Aggarwal, B. B. (2006). Resveratrol inhibits proliferation, induces apoptosis, and overcomes chemoresistance through down-regulation of STAT3 and nuclear factor-κB–regulated antiapoptotic and cell survival gene products in human multiple myeloma cells. *Blood 109*: 2293–2302.

Bian, Y., Wang, X., Zheng, Z., Ren, G., Zhu, H., Qiao, M. & Li, G. (2022). Resveratrol drives cancer cell senescence via enhancing p38MAPK and DLC1 expressions. *Food Funct 13*: 3283–3293.

Bijangi-Vishehsaraei, K., Reza Saadatzadeh, M., Wang, H., Nguyen, A., Kamocka, M. M., Cai, W., Cohen-Gadol, A. A., Halum, S. L., Sarkaria, J. N., Pollok, K. E. & Safa, A. R. (2017). Sulforaphane suppresses the growth of glioblastoma cells, glioblastoma stem cell-like spheroids, and tumor xenografts through multiple cell signaling pathways. *J Neurosurg 127*: 1219–1230.

Chai, R., Fu, H., Zheng, Z., Liu, T., Ji, S. & Li, G. (2017). Resveratrol inhibits proliferation and migration through SIRT1 mediated post-translational modification of PI3K/AKT signaling in hepatocellular carcinoma cells. *Molecular Medicine Reports 16*: 8037–8044.

Chen, C. C., Huang, J. S., Wang, T. H., Kuo, C. H., Wang, C. J., Wang, S. H. & Leu, Y. L. (2017). Dihydrocoumarin, an HDAC inhibitor, increases DNA damage sensitivity by inhibiting Rad52. *International Journal of Molecular Sciences 18*: 2655.

Chen, C.-T., Hsieh, M.-J., Hsieh, Y.-H., Hsin, M.-C., Chuang, Y.-T., Yang, S.-F., Yang, J.-S. & Lin, C.-W. (2018). Sulforaphane suppresses oral cancer cell migration by regulating cathepsin S expression. *Oncotarget 9*: 17564–17575.

Cheng, Y. M., Tsai, C. C. & Hsu, Y. C. (2016). Sulforaphane, a dietary isothiocyanate, induces G_2/M arrest in cervical cancer cells through cyclinB1 downregulation and GADD45β/ CDC2 association. *International Journal of Molecular Sciences 17*: 1530.

Chiou, Y. S., Lan, Y. M., Lee, P. S., Lin, Q., Nagabhushanam, K., Ho, C. T. & Pan, M. H. (2022). Piceatannol prevents colon cancer progression via dual-targeting to M2-polarized tumor-associated macrophages and the TGF-β1 positive feedback signaling pathway. *Mol Nutr Food Res*: e2200248.

Dickinson, S. E., Rusche, J. J., Bec, S. L., Horn, D. J., Janda, J., Rim, S. H., Smith, C. L. & Bowden, G. T. (2015). The effect of sulforaphane on histone deacetylase activity in keratinocytes: Differences between in vitro and in vivo analyses. *Mol Carcinog 54*: 1513–1520.

Ding, D., Wei, S., Song, Y., Li, L., Du, G., Zhan, H. & Cao, Y. (2013). Osthole exhibits anti-cancer property in rat glioma cells through inhibiting PI3K/Akt and MAPK signaling pathways. *Cellular Physiology and Biochemistry 32*: 1751–1760.

Druesne, N., Pagniez, A., Mayeur, C., Thomas, M., Cherbuy, C., Duée, P. H., Martel, P. & Chaumontet, C. (2004). Diallyl disulfide (DADS) increases histone acetylation and p21(waf1/cip1) expression in human colon tumor cell lines. *Carcinogenesis 25*: 1227–1236.

Du, M., Zhang, Z. & Gao, T. (2017). Piceatannol induced apoptosis through up-regulation of microRNA-181a in melanoma cells. *Biological Research 50*: 36.

Ferrari, I. (2022). Potential garlic-derived compound S-Allyl-mercapto-L-cysteine (SAMC) for prevention cancer. DOI 10.13140/RG.2.2.12752.58880.

Filomeni, G., Graziani, I., Rotilio, G. & Ciriolo, M. R. (2007). Trans-resveratrol induces apoptosis in human breast cancer cells MCF-7 by the activation of MAP kinases pathways. *Genes & Nutrition 2*: 295–305.

Fung, K. Y. C., Brierley, G. V., Henderson, S., Hoffmann, P., McColl, S. R., Lockett, T., Head, R. & Cosgrove, L. (2011). Butyrate-induced apoptosis in HCT116 colorectal cancer cells includes induction of a cell stress response. *Journal of Proteome Research 10*: 1860–1869.

Ge, J., Liu, Y., Li, Q., Guo, X., Gu, L., Ma, Z. G. & Zhu, Y. P. (2013). Resveratrol induces apoptosis and autophagy in t-cell acute lymphoblastic leukemia cells by inhibiting Akt/mTOR and activating p38-MAPK. *Biomedical and Environmental Sciences 26*: 902–911.

Geng, H.-W., Yin, F.-Y., Zhang, Z.-F., Gong, X. & Yang, Y. (2021). Butyrate suppresses glucose metabolism of colorectal cancer cells via GPR109a-AKT signaling pathway and enhances chemotherapy. *Frontiers in Molecular Biosciences 8*: 634874.

Greco, G., Schnekenburger, M., Catanzaro, E., Turrini, E., Ferrini, F., Sestili, P., Diederich, M. & Fimognari, C. (2021). Discovery of sulforaphane as an inducer of ferroptosis in U-937 leukemia cells: expanding its anticancer potential. *Cancers (Basel) 14*: 76.

Han, R., Nusbaum, O., Chen, X. & Zhu, Y. (2020). Valeric acid suppresses liver cancer development by acting as a novel HDAC inhibitor. *Molecular Therapy – Oncolytics 19*: 8–18.

Heckmann, B. L., Yang, X., Zhang, X. & Liu, J. (2013). The autophagic inhibitor 3-methyladenine potently stimulates PKA-dependent lipolysis in adipocytes. *Br J Pharmacol 168*: 163–171.

Hu, H., Zhang, X. P., Wang, Y. L., Chua, C. W., Luk, S. U., Wong, Y. C., Ling, M. T., Wang,
X. F. & Xu, K. X. (2011). Identification of a novel function of Id-1 in mediating the
anticancer responses of SAMC, a water-soluble garlic derivative, in human bladder
cancer cells. *Molecular Medicine Reports 4*: 9–16.

Hu, W. H., Dai, D. K., Zheng, B. Z., Duan, R., Dong, T. T., Qin, Q. W. & Tsim, K. W.
(2020). Piceatannol, a natural analog of resveratrol, exerts anti-angiogenic efficiencies
by blockage of vascular endothelial growth factor binding to its receptor. *Molecules
25*: 3769.

Jarząb, A., Grabarska, A., KiełBus, M., Jeleniewicz, W., DmoszyńSka-Graniczka, M., Skalicka-
WoźNiak, K., Sieniawska, E., Polberg, K. & Stepulak, A. (2014). Osthole induces apoptosis, suppresses cell-cycle progression and proliferation of cancer cells. *Anticancer Res
34*: 6473.

Jeon, Y. K., Yoo, D. R., Jang, Y. H., Jang, S. Y. & Nam, M. J. (2011). Sulforaphane induces
apoptosis in human hepatic cancer cells through inhibition of 6-phosphofructo-
2-kinase/fructose-2,6-biphosphatase4, mediated by hypoxia inducible factor-1-
dependent pathway. *Biochimica et Biophysica Acta (BBA) - Proteins and Proteomics
1814*: 1340–1348.

Ji, S., Zheng, Z., Liu, S., Ren, G., Gao, J., Zhang, Y. & Li, G. (2018). Resveratrol promotes
oxidative stress to drive DLC1 mediated cellular senescence in cancer cells. *Exp Cell
Res 370*: 292–302.

Jia, H.-w., Wu, Y., Li, R.-x., Zhao, H.-c., Wang, G.-j. & Sun, J.-f. (2019). Resveratrol induces
apoptosis, autophagy and endoplasmic reticulum stress in colon cancer cells. *Clinical
Surgery Research Communications 3*: 19–28.

Jiang, X.-y., Zhu, X.-s., Xu, H.-y., Zhao, Z.-x., Li, S.-y., Li, S.-z., Cai, J.-h. & Cao, J.-m.
(2017). Diallyl trisulfide suppresses tumor growth through the attenuation of Nrf2/Akt
and activation of p38/JNK and potentiates cisplatin efficacy in gastric cancer treatment.
Acta Pharmacologica Sinica 38: 1048–1058.

Jo, G. H., Kim, G.-Y., Kim, W.-J., Park, K. Y. & Choi, Y. H. (2014). Sulforaphane induces
apoptosis in T24 human urinary bladder cancer cells through a reactive oxygen species-
mediated mitochondrial pathway: the involvement of endoplasmic reticulum stress and
the Nrf2 signaling pathway. *Int J Oncol 45*: 1497–1506.

Kim, D., Lee, I. S., Jung, J. H. & Yang, S. I. (1999). Psammaplin A, a natural bromotyrosine
derivative from a sponge, possesses the antibacterial activity against methicillin-
resistant staphylococcus aureus and the DNA gyrase-inhibitory activity. *Arch Pharm
Res 22*: 25–29.

Kim, D. H., Shin, J. & Kwon, H. J. (2007). Psammaplin A is a natural prodrug that inhibits
class I histone deacetylase. *Exp Mol Med 39*: 47–55.

Kim, H. J., Kim, T. H., Seo, W. S., Yoo, S. D., Kim, I. H., Joo, S. H., Shin, S., Park, E. S., Ma,
E. S. & Shin, B. S. (2012). Pharmacokinetics and tissue distribution of psammaplin A, a
novel anticancer agent, in mice. *Arch Pharm Res 35*: 1849–1854.

Kim, S.-E., Shin, S.-H., Lee, J.-Y., Kim, C.-H., Chung, I.-K., Kang, H.-M., Park, H.-R., Park,
B.-S. & Kim, I.-R. (2018). Resveratrol induces mitochondrial apoptosis and inhibits
epithelial-mesenchymal transition in oral squamous cell carcinoma cells. *Nutr Cancer
70*: 125–135.

Kimura, Y. (2022). Long-term oral administration of piceatannol (3,5,3',4'-tetrahydroxystilbene)
attenuates colon tumor growth induced by azoxymethane plus dextran sulfate sodium in
C57BL/6J mice. *Nutr Cancer 74*: 2184–2195.

Koolivand, M., Ansari, M., Moein, S., Afsa, M. & Malekzadeh, K. (2022). The inhibitory effect
of sulforaphane on the proliferation of acute myeloid leukemia cell lines through controlling miR-181a. *Cell J 24*: 44–50.

Koushki, M., Amiri-Dashatan, N., Ahmadi, N., Abbaszadeh, H. A. & Rezaei-Tavirani, M. (2018). Resveratrol: a miraculous natural compound for diseases treatment. *Food Sci Nutr 6*: 2473–2490.

Krishnan, M., Saraswathy, S., Singh, S., Saggu, G. K., Kalra, N., Agrawala, P. K., Abraham, K. M. & Das Toora, B. (2022). Sulforaphane inhibits histone deacetylase causing cell cycle arrest and apoptosis in oral squamous carcinoma cells. *Medical Journal Armed Forces India.*

Lang, F., Qin, Z., Li, F., Zhang, H., Fang, Z. & Hao, E. (2015). Apoptotic cell death induced by resveratrol is partially mediated by the autophagy pathway in human ovarian cancer cells. *PLoS One 10*: e0129196.

Lea, M. A., Rasheed, M., Randolph, V. M., Khan, F., Shareef, A. & desBordes, C. (2002). Induction of histone acetylation and inhibition of growth of mouse erythroleukemia cells by S-allylmercaptocysteine. *Nutr Cancer 43*: 90–102.

Li, N., Cui, T., Guo, W., Wang, D. & Mao, L. (2019). MiR-155-5p accelerates the metastasis of cervical cancer cell via targeting TP53INP1. *OncoTargets and therapy 12*: 3181–3196.

Li, Y., Dong, J., Xiao, H., Zhang, S., Wang, B., Cui, M. & Fan, S. (2020). Gut commensal derived-valeric acid protects against radiation injuries. *Gut Microbes 11*: 789–806.

Li, Y., Zhang, T., Korkaya, H., Liu, S., Lee, H. F., Newman, B., Yu, Y., Clouthier, S. G., Schwartz, S. J., Wicha, M. S. & Sun, D. (2010). Sulforaphane, a dietary component of broccoli/broccoli sprouts, inhibits breast cancer stem cells. *Clin Cancer Res 16*: 2580–2590.

Liao, X. X., Dai, Y. Z., Zhao, Y. Z. & Nie, K. (2022). Gasdermin E: a prospective target for therapy of diseases. *Front Pharmacol 13*: 855828.

Liu, R., Yang, Y. N., Yi, L., Qing, J., Li, Q. Y., Wang, W. S., Wang, J., Tang, Y. X. & Tan, H. (2018). Diallyl disulfide effect on the invasion and migration ability of HL-60 cells with a high expression of DJ-1 in the nucleus through the suppression of the Src signaling pathway. *Oncology Letters 15*: 6377–6385.

Liu, Y., Yan, J., Han, X. & Hu, W. (2015). Garlic-derived compound S-allylmercaptocystcine (SAMC) is active against anaplastic thyroid cancer cell line 8305C (HPACC). *Technol Health Care 23 Suppl 1*: S89–93.

Lv, Y., So, K. F., Wong, N. K. & Xiao, J. (2019). Anti-cancer activities of S-allylmercaptocysteine from aged garlic. *Chin J Nat Med 17*: 43–49.

Miki, H., Uehara, N., Kimura, A., Sasaki, T., Yuri, T., Yoshizawa, K. & Tsubura, A. (2012). Resveratrol induces apoptosis via ROS-triggered autophagy in human colon cancer cells. *Int J Oncol 40*: 1020–1028.

Navarro-Orcajada, S., Conesa Valverde, I., Vidal-Sánchez, F., Matencio, A., Albaladejo-Maricó, L., García-Carmona, F. & López-Nicolás, J. (2022). Stilbenes: characterization, bioactivity, encapsulation and structural modifications. A review of their current limitations and promising approaches. *Crit Rev Food Sci Nutr*: 1–19.

Nian, H., Delage, B., Pinto, J. T. & Dashwood, R. H. (2008). Allyl mercaptan, a garlic-derived organosulfur compound, inhibits histone deacetylase and enhances Sp3 binding on the P21WAF1 promoter. *Carcinogenesis 29*: 1816–1824.

Osuka, S. & Van Meir, E. G. (2017). Overcoming therapeutic resistance in glioblastoma: the way forward. *J Clin Invest 127*: 415–426.

Panebianco, C., Villani, A., Pisati, F., Orsenigo, F., Ulaszewska, M., Latiano, T. P., Potenza, A., Andolfo, A., Terracciano, F., Tripodo, C., Perri, F. & Pazienza, V. (2022). Butyrate, a postbiotic of intestinal bacteria, affects pancreatic cancer and gemcitabine response in in vitro and in vivo models. *Biomedicine & Pharmacotherapy 151*: 113163.

Peng, K. Y. & Chou, T. C. (2022). Osthole exerts inhibitory effects on hypoxic colon cancer cells via EIF2 phosphorylation-mediated apoptosis and regulation of HIF-1. *Am J Chin Med 50*: 621–637.

Pey, A. L., Megarity, C. F. & Timson, D. J. (2019). NAD(P)H quinone oxidoreductase (NQO1): an enzyme which needs just enough mobility, in just the right places. *Biosci Rep 39*.

Pledgie-Tracy, A., Sobolewski, M. D. & Davidson, N. E. (2007). Sulforaphane induces cell type-specific apoptosis in human breast cancer cell lines. *Mol Cancer Ther 6*: 1013–1021.

Puccinelli, M. T. & Stan, S. D. (2017). Dietary bioactive diallyl trisulfide in cancer prevention and treatment. *International Journal of Molecular Sciences 18*: 645.

Puttalingaiah, S., Anantharaju, P., Veeresh, P., Dey, S., Bovilla, V. & Madhunapantula, S. (2017). Diallyl disulfide (DADS) retards the growth of breast cancer cells in vitro and in vivo through apoptosis induction. *Biomedical and Pharmacology Journal 10*: 1619–1630.

Seki, T., Hosono, T., Hosono-Fukao, T., Inada, K., Tanaka, R., Ogihara, J. & Ariga, T. (2008). Anticancer effects of diallyl trisulfide derived from garlic. *Asia Pac J Clin Nutr 17 Suppl 1*: 249–252.

Shan, Y., Wei, Z., Tao, L., Wang, S., Zhang, F., Shen, C., Wu, H., Liu, Z., Zhu, P., Wang, A., Chen, W. & Lu, Y. (2016). Prophylaxis of diallyl disulfide on skin carcinogenic model via p21-dependent Nrf2 stabilization. *Scientific Reports 6*: 35676.

Sharifi-Rad, J., Cruz-Martins, N., López-Jornet, P., Lopez, E. P.-F., Harun, N., Yeskaliyeva, B., Beyatli, A., Sytar, O., Shaheen, S., Sharopov, F., Taheri, Y., Docea, A. O., Calina, D. & Cho, W. C. (2021). Natural coumarins: exploring the pharmacological complexity and underlying molecular mechanisms. *Oxid Med Cell Longev 2021*: 6492346.

Shi, F., Li, Y., Han, R., Fu, A., Wang, R., Nusbaum, O., Qin, Q., Chen, X., Hou, L. & Zhu, Y. (2021). Valerian and valeric acid inhibit growth of breast cancer cells possibly by mediating epigenetic modifications. *Scientific Reports 11*: 2519.

Shih, A., Davis, F. B., Lin, H.-Y. & Davis, P. J. (2002). Resveratrol induces apoptosis in thyroid cancer cell lines via a MAPK- and p53-dependent mechanism. *Journal of Clinical Endocrinology & Metabolism 87*: 1223–1232.

Shim, J. S., Lee, H. S., Shin, J. & Kwon, H. J. (2004). Psammaplin A, a marine natural product, inhibits aminopeptidase N and suppresses angiogenesis in vitro. *Cancer Lett 203*: 163–169.

Shirin, H., Pinto, J. T., Kawabata, Y., Soh, J.-W., Delohery, T., Moss, S. F., Murty, V., Rivlin, R. S., Holt, P. R. & Weinstein, I. B. (2001). Antiproliferative effects of S-allylmercaptocysteine on colon cancer cells when tested alone or in combination with sulindac sulfide1. *Cancer Res 61*: 725–731.

Sirerol, J. A., Rodríguez, M. L., Mena, S., Asensi, M. A., Estrela, J. M. & Ortega, A. L. (2016). Role of natural stilbenes in the prevention of cancer. *Oxid Med Cell Longev 2016*: 3128951.

Sita, G., Graziosi, A., Hrelia, P. & Morroni, F. (2021). Sulforaphane causes cell cycle arrest and apoptosis in human glioblastoma U87MG and U373MG cell lines under hypoxic conditions. *International Journal of Molecular Sciences 22*: 11201.

Son, I. H., Lee, S. I., Yang, H. D. & Moon, H. I. (2007). Bis(4-hydroxybenzyl)sulfide: a sulfur compound inhibitor of histone deacetylase isolated from root extract of Pleuropterus ciliinervis. *Molecules 12*: 815–820.

Su, N., Li, L., Zhou, E., Li, H., Wu, S. & Cao, Z. (2022). Resveratrol downregulates miR-155–5p to block the malignant behavior of gastric cancer cells. *Biomed Res Int 2022*: 6968641.

Suppipat, K., Park, C. S., Shen, Y., Zhu, X. & Lacorazza, H. D. (2012). Sulforaphane induces cell cycle arrest and apoptosis in acute lymphoblastic leukemia cells. *PLoS One 7*: e51251.

Wang, H., Sun, N., Li, X., Li, K., Tian, J. & Li, J. (2016). Diallyl trisulfide induces osteosarcoma cell apoptosis through reactive oxygen species-mediated downregulation of the
PI3K/Akt pathway. *Oncol Rep 35*: 3648–3658.

Wang, J., Huangfu, M., Li, X., Han, M., Liu, G., Yu, D., Zhou, L., Dou, T., Liu, Y., Guan,
X., Wei, R. & Chen, X. (2022). Osthole induces apoptosis and caspase-3/GSDME-
dependent pyroptosis via NQO1-mediated ROS generation in HeLa cells. *Oxid Med
Cell Longev 2022*: 8585598.

Wee, C., Kim, J., Kim, H., Kang, H.-C., Suh, S., Shin, B., Ma, E. & Il Han, K. (2019).
Psammaplin A-modified novel radiosensitizers for human lung cancer and glioblastoma
cells. *Journal of Radiation Protection and Research 44*: 15–25.

Wei, W., Tang, C., Zhan, X., Yi, H. & Li, C. (2013). Effect of DJ-1 siRNA on biological
behavior of human lung squamous carcinoma SK-MES-1 cells. *Zhong Nan Da Xue Xue
Bao Yi Xue Ban 38*: 7–13.

Witaicenis, A., Seito, L.N., Di Stasi, L.C. (2010) Intestinal anti-inflammatory activity of
esculetin and 4-methylesculetin in the trinitrobenzenesulphonic acid model of rat col-
itis. *Chem Biol Interact 186*(2): 211–218. doi: 10.1016/j.cbi.2010.03.045.

Wu, F., Liang, Y. Q. & Huang, Z. M. (2009). The expression of DJ-1 gene in human hepatocellular
carcinoma and its relationship with tumor invasion and metastasis. Zhonghua Gan Zang
Bing Za Zhi *17*: 203–206.

Xu, S., Liu, C.-X., Xu, W., Huang, L., Zhao, J.-Y. & Zhao, S.-M. (2017). Butyrate induces
apoptosis by activating PDC and inhibiting complex I through SIRT3 inactivation.
Signal Transduction and Targeted Therapy 2: 16035.

Yang, D., Gu, T., Wang, T., Tang, Q. & Ma, C. (2010). Effects of osthole on migration and invasion in breast cancer cells. *Biosci Biotechnol Biochem 74*: 1430–1434.

Yi, L. & Su, Q. (2013). Molecular mechanisms for the anti-cancer effects of diallyl disulfide.
Food Chem Toxicol 57: 362–370.

Yin, L., Xiao, X., Georgikou, C., Luo, Y., Liu, L., Gladkich, J., Gross, W. & Herr, I. (2019).
Sulforaphane induces miR135b-5p and its target gene, RASAL2, thereby inhibiting the
progression of pancreatic cancer. *Molecular Therapy - Oncolytics 14*: 74–81.

Yin, X., Zhang, R., Feng, C., Zhang, J., Liu, D., Xu, K., Wang, X., Zhang, S., Li, Z., Liu, X. &
Ma, H. (2014). Diallyl disulfide induces G2/M arrest and promotes apoptosis through
the p53/p21 and MEK-ERK pathways in human esophageal squamous cell carcinoma.
Oncol Rep 32: 1748–1756.

Yue, Z., Guan, X., Chao, R., Huang, C., Li, D., Yang, P., Liu, S., Hasegawa, T., Guo, J. & Li,
M. (2019). Diallyl disulfide induces apoptosis and autophagy in human osteosarcoma
MG-63 cells through the PI3K/Akt/mTOR pathway. *Molecules 24*: 2665.

Zhang, G., Xu, Y. & Zhou, H. F. (2019). Esculetin inhibits proliferation, invasion, and migration of laryngeal cancer in vitro and in vivo by inhibiting janus kinas (JAK)-signal
transducer and activator of transcription-3 (STAT3) activation. *Medical Science
Monitor: International Medical Journal of Experimental and Clinical Research
25*: 7853–7863.

Zhang, H., Wang, K., Lin, G. & Zhao, Z. (2014). Antitumor mechanisms of S-allyl
mercaptocysteine for breast cancer therapy. *BMC Complementary and Alternative
Medicine 14*: 270.

Zhang, H. Y., Wang, H. Q., Liu, H. M., Guan, Y. & Du, Z. X. (2008). Regulation of tumor
necrosis factor-related apoptosis-inducing ligand-induced apoptosis by DJ-1 in thyroid
cancer cells. *Endocr Relat Cancer 15*: 535–544.

Zhang, Y., Li, H. Y., Zhang, Z. H., Bian, H. L. & Lin, G. (2014). Garlic-derived compound
S-allylmercaptocysteine inhibits cell growth and induces apoptosis via the JNK and p38
pathways in human colorectal carcinoma cells. *Oncology Letters 8*: 2591–2596.

Zhang, Y., Lu, Q., Li, N., Xu, M., Miyamoto, T. & Liu, J. (2022). Sulforaphane suppresses metastasis of triple-negative breast cancer cells by targeting the RAF/MEK/ERK pathway. *npj Breast Cancer 8*: 40.

Zhang, Z., Li, C., Shang, L., Zhang, Y., Zou, R., Zhan, Y. & Bi, B. (2016). Sulforaphane induces apoptosis and inhibits invasion in U251MG glioblastoma cells. *SpringerPlus 5*: 235.

Zhao, Z., Bo, Z., Gong, W. & Guo, Y. (2020). Inhibitor of differentiation 1 (Id1) in cancer and cancer therapy. *International Journal of Medical Sciences 17*: 995–1005.

Zhou, J. H., Cheng, H. Y., Yu, Z. Q., He, D. W., Pan, Z. & Yang, D. T. (2011). Resveratrol induces apoptosis in pancreatic cancer cells. *Chin Med J (Engl) 124*: 1695–1699.

7 Flavonoid Inhibitors of Histone Deacetylases in Concert with Conventional Chemotherapeutic Agents for Phenomenal Therapy Against Cancer

Neoplastic cells frequently develop resistance when they are subjected to single therapeutics. Apart from this, singlet therapy often proves fruitful at higher concentrations thereby inducing substantial side-effects on typical body cells. Flavonoid histone deacetylase inhibitors (HDACi), like other therapeutic agents when used in monotherapy of cancer, often fail to yield preferred preclinical and clinical outcome. As these inhibitors as single agents require a higher concentration to induce pharmacological effect in cancer models, these concentrations often sensitize normal body cells to cytotoxic effects of HDACi culminating in debilitating toxicity. This agonizing concern has been resolved by using these flavone HDACi jointly with standard anticancer agents. This tactic on the one hand produce expected therapeutic benefit either in synergistic or additive forms even at doses that are not sufficient to elicit overt toxicity towards usual cells. Moreover, the combined therapy strongly alleviates drug resistance, the premier disquiet in anticancer therapy. Taking all this into consideration, this chapter will entirely cynosure on combinatorial therapy involving flavone HDACi from plant sources cooperatively with regular anticancer molecules.

Combinatorial therapy, the cornerstone of anticancer therapy, makes use of two or greater number of drugs that are used either sequentially or concurrently on disease models. This approach offers multiple benefits over monotherapy, including better efficacy, low toxicity and the lesser likelihood of drug resistance (Bayat Mokhtari et al. 2017). Histone deacetylase inhibitors (HDACi) of the flavonoid type have shown significant therapeutic advantages, which are further improved when these inhibitors are used along with normal therapeutics (Ganai, Sheikh & Baba 2021).

DOI: 10.1201/9781003294863-7

Flavonoid-type HDACi, when used in combined fashion with conventional agents, raises their anticancer activity (Yan et al. 2020). For making this chapter eye-catching, the most well-studied members of different flavonoid groups in the context of combined therapy against bellicose malignancies will be discussed.

7.1 FLAVONES IN ASSOCIATION WITH STANDARD AGENTS

A detailed account of flavones which come within the confines of flavonoids has already been given in Chapter 5 dealing with flavonoids. Luteolin, apigenin and chrysin are some members of the flavone group that have been highly studied as single agents as well as in association with other usual therapeutics (Khan et al. 2021).

7.1.1 CHRYSIN IN COOPERATION WITH STANDARD CHEMOTHERAPEUTICS

It has been verified through experiments that cancer cells develop resistance against cisplatin, the conventional molecule, and this significantly hampers its clinical applications (Ganai et al. 2021). Chrysin and cisplatin when used in combination markedly enhanced HepG2 apoptosis. This combination increased p53 phosphorylation and its amassing by stimulating ERK1/2 in these cells. This in turn favoured down-regulation of Bcl-2 (antiapoptotic protein) and concurrent up-regulation of death receptor 5 (DR5) and Bax (pro-apoptotic proteins). Dual treatment induced apoptosis of these cells through the intrinsic and extrinsic apoptotic pathway. While the intrinsic mechanism was evident from cytochrome c release and caspase-9 activation, the extrinsic pathway was confirmed through caspase-8 activation (Li et al. 2015). Chrysin-cisplatin union as indicated by the results is a stunning therapeutic option for circumventing cisplatin resistance. Gastric cancer is one among the many cancer types with high mortality and 5-fluorouracil (5-FU) is considered as a high standard drug for subduing this malignancy. However, its efficacy is hampered due to resistance mechanisms generated by these cells. Chrysin was administered along with 5-FU in the human gastric adenocarcinoma cell line (AGS) and its 5-FU resistant version (AGS/FR). Significantly improved cytotoxic effects over single agent therapy were noticed when the combination of the aforesaid molecules was used. The cytotoxic effects of 5-FU were elevated in sensitive and resistant gastric cancer lines were strengthened by chrysin through induction of cell cycle arrest (Lee, Lee & Jung 2021).

For circumventing colorectal cancer, 5-FU has great importance. Against this cancer 5-FU was investigated in conjunction with flavone chrysin. In mice CRC was induced through combined use of azoxymethane and dextran sodium sulfate. Aberrant crypt foci (ACF) are the most premier neoplastic areas (lesions), which on advancement transform into polyp (Alrawi et al. 2006). In other words ACF can be regarded as colorectal cancer precursors (Cheng & Lai 2003). The effects of 5-FU were boosted once it was used jointly with chrysin. The chrysin-5FU combination resulted in additional down-regulation of COX-2 when compared to cells treated with individual agents. Thus chrysin may be used as an adjunct with 5-FU to potentiate the cytotoxic effects of the latter (Yosefi et al. 2022).

7.1.2 LUTEOLIN-CONVENTIONAL DRUGS COMBINATION

Luteolin has manifested encouraging results in a broad range of malignancies such as breast, lung, glioma, pancreatic, prostate, gastric cancer and many more. Like chrysin, its elevated therapeutics benefits have been achieved only when it has been used with other anticancer agents like gemicitabine, doxorubicin, cisplatin, paclitaxel and many more (Ganai et al. 2021). Paclitaxel is used as the predominant therapy for many cancers, but its applications are limited due to significant side-effects and the resistance generated by cancer cells towards it. The combination of this drug with luteolin showed synergistic effect in inhibiting proliferation, epithelial to mesenchymal transformation, migration and cytotoxicity of oesophageal cancer cells. This synergistic effect in obstructing the growth of tumour cells also occurred under conditions of *in vivo* and no noticeable toxicity was observed. These effects have been ascribed to inhibition of Class III HDAC namely SIR1 and modulation of ROS by this combination in these cells (Qin et al. 2021). Enhanced cytotoxic effect was recorded when flavone luteolin was used in combination with taxol against breast cancer cells (Tsai et al. 2021).

Luteolin has also been evaluated against *in vivo* pancreatic cancer models in cooperation with gemcitabine. The defined molecules were tested separately as well as combinedly for six weeks against an orthotopic mouse model. In comparison to the control the luteolin-gemcitabine combination induced a substantial depletion of tumour mass. This was accompanied with a marked elevation in cytochrome c and caspase 3. On the other hand, GSK-3β, K-ras and the ratio of bcl-2/bax underwent considerable decline (Johnson et al. 2015). Ovarian cells, like other cancer cells, also develop resistance to cisplatin. Luteolin was tested along with cisplatin in an ovarian cancer line resistant to cisplatin, in addition to a xenotransplant mice model. In CAOV3/DDP cells resistant to cisplatin, luteolin enhanced the antiproliferative action of cisplatin. The apoptosis of these cells was increased when these cells were given dual treatment. Cisplatin induced Bcl-2 down-regulation, which was further increased by the addition of luteolin. Apart from this, the mentioned combination evinced synergy in alleviating migration and invasion of resistant cells. Further, additional decline in *in vivo* tumour growth occurred, induced by cisplatin in combination with luteolin (Wang et al. 2018).

Another study explored that luteolin sensitization of cancer cells to cisplatin is dependent on p53. This crux is derived as luteolin was unable to sensitize p53 mutant and p53 knockdown cells to cisplatin. Luteolin up-regulated p53 only at the protein level and not at the transcript level. Further investigation revealed that luteolin up-regulates p53 by inducing its phosphorylation through activations of c-Jun NH$_2$-terminal kinase (JNK). The phosphorylation of p53 hampers its ubiquitination and eventual clearance through proteasomes resulting in its stabilization. This effect was also observed in the xenograft model confirming that luteolin stabilizes p53 both under *in vitro* and *in vivo* conditions (Shi et al. 2007). Conclusively, luteolin heightens the cytotoxic effect of cisplatin through JNK activation, which in turn results in phosphorylation of p53 and its subsequent stabilization. Just recently, luteolin has been assessed against colon cancer in association with curcumin. A low-dose combination of these phytochemicals suppressed the growth of cultured colon cancer cells in a synergistic manner. A similar result was noticed in the xenograft mice model (Aromokeye & Si 2022). These results indicate that consuming foods rich in luteolin and curcumin may offer therapeutic benefit by inhibiting the growth of abnormal colon cancer cells.

7.1.3 APIGENIN IN COLLABORATION WITH STANDARD CANCER THERAPEUTICS

Apigenin has demonstrated significant antiproliferative activity against a wide range of cancers. Its anticancer effect is more profound when this flavone is used along with other anticancer molecules (Ganai 2017; Singh et al. 2022). The apigenin and curcumin combination showed synergistic effect in inducing apoptosis in lung cancer cells. This combination arrested these cells at the G_2/M phase and induced potent depolymerisation of microtubules at interphase (Choudhury et al. 2013). The luteolin–doxorubicin combination has been evaluated on prostate cancer cells (androgen insensitive). Co-treatment with these agents induced substantial cytoxicity in these cells through up-regulation of caspases and Bax. Cell cycle arrest was also elicited by this combination that has been attributed to increased expression of p27 and p21. The blockade of cell migration was also provoked by co-administration of these agents by inhibiting the Twist, MMPs and Snail. While apigenin caused Akt dephosphorylation, it substantially reduced phosphorylation of ERK and last, but no way least, elevated expression of PTEN. Moreover, the apigenin-doxorubicin combination proved to be more effective over individual interventions in obstructing the phosphorylation of AKT and PI3K (Ayyildiz et al. 2021).

The major impediment in hepatocellular carcinoma therapy is chemo-resistance. Apigenin has been studied in the context of sensitizing hepatocellular carcinoma cells (doxorubicin resistant) to doxorubicin. Flavone apigenin sensitized these cells to doxorubicin through up-regulation of miR-101. The overexpression of this miR reproduced the effect of apigenin in these cells, clearly indicating that the apigenin-induced effect of sensitizing BEL-7402/ADM cells is mediated by miR-101. Most importantly it was delineated that this miR targets 3′-UTR of transcription factor nuclear factor erythroid 2–related factor 2 (Nrf2) (Gao, Zhang & Ke 2017). Thus, apigenin induces chemosensitivity in doxorubicin-resistant cells of hepatocellular carcinoma through induction of miR-101 and subsequent modulation of Nrf2. Apigenin and paclitaxel even at low doses induced proapoptotic effects in cancer cells in a synergistic manner. This combination caused amassing of superoxides in cervical cancer cells. This dual treatment provoked apoptosis in these cells by enhancing the ROS production. Apigenin-induced ROS elevation was due to the inhibition of the enzymatic function of superoxide dismutase (SOD). Further studies demonstrated that caspase-2 also has cross-talk with apoptosis triggered by apigenin-paclitaxel (Xu et al. 2011). On the whole, these results suggest that apigenin sensitizes HeLa cells to cytotoxic effects of paclitaxel by inhibiting SOD function resulting in ROS accumulation and subsequent caspase-2 cleavage. Bruton tyrosine kinase (BTK) is considered a critical enzyme for its role in B cell development and maturation. The inhibitor of this kinase, namely abivertinib, has been tested on diffuse large B cell lymphoma, the malignancy of B lymphocytes. Apigenin elicited proliferation inhibition, blocked cell cycle and stimulated apoptosis in DLBCL cells through modulation of proteins involved in cell cycle, BCL-XL down-regulation and through caspase activation. Apigenin reduced the survival of these cells by inhibiting the PI3K/mTOR pathway (Huang et al. 2020). Combining this flavone with abivertinib synergistically induced apoptosis in these cells by obstructing p-GSK3-β, which in turn modulates its downstream molecular targets.

7.1.4 Rutin in Combinatorial Therapy

Like apigenin, luteolin and chrysin, rutin has manifested antiproliferative activity against a panel of cancers. As expected, it has proved stupendous in combination with conventional drugs. The combination of rutin with cisplatin substantially extended the life span of mice bearing Dalton's lymphoma. Individual intervention with cisplatin induced genotoxic effects in the above-mentioned mice model, and these effects were soothed when apigenin was added to cisplatin treatment (Prasad & Prasad 2021b). Another evidence-based study revealed that cisplatin-induced neurotoxic effects are pacified by using rutin jointly with cisplatin. The expression of certain genes that were altered by cisplatin was reinstated to their normal state by this combined regime. This neuroprotective effect of rutin has been credited to its antioxidant potential (Alshazi et al. 2017). In tumour-carrying mice, cisplatin induced histotoxicity. This drug caused damage to kidneys, liver and testes. When rutin was added to cisplatin, the histotoxicity of cisplatin was mitigated. From these findings it is obvious that rutin has histoprotective effect and can be used as an adjunct with standard drugs for ameliorating the off-target effects of the latter (Prasad & Prasad 2021a).

7.2 ISOFLAVONES AND CONVENTIONAL DRUGS COMBINATION

Among this group of flavonoids, genistein and daidzein are most well studied. Taking this into consideration, only these two isoflavones will be discussed under this heading (Goh et al. 2022).

7.2.1 Standard Drugs in Combination with Genistein

Genistein, the isoflavone, improved the pharmokinetic properties of paclitaxel in rat models (Li & Choi 2007). This isoflavone has the potential to restrain the growth of multiple cancers. The impact of genistein in association with cisplatin was explored in human medulloblastoma cells. In the HTB-186 cell line, the addition of genistein enhanced the cisplatin-induced growth inhibitory effect by 2.8 fold. The antiproliferative effect elicited by cisplatin was elevated to 1.3 fold on combination with genistein. The effects exerted by genistein–cisplatin were premierly synergistic. However, genistein association with vincristine elevated the antiproliferative effect of the latter only to a minimal to moderate degree (Khoshyomn et al. 2000). Genistein–cisplatin duo was also investigated in cervical cancer cells, namely CaSki, and it was observed that genistein enhanced the cytotoxic effects of cisplatin in these cells. In comparison to the control, genistein–cisplatin declined the p-ERK1/2 level by 37% and, on the other hand, elevated p53 by 304%. Further, the dual treatment enhanced the caspase-3 cleavage by 115% and lessened the Bcl2 level by 69% (Liu et al. 2019). Thus, genistein-augmented cisplatin invoked effects in CaSki cells by modulating a variety of molecular targets. This combination was also tested on the hepatocellular carcinoma cell line, namely HCCLM3, and it was concluded that the duo induced synergistic effect in these cells. Genistein lowered the elevated expression of MMP-2-induced by cisplatin (Chen et al. 2013). These results suggest that the cisplatin-stimulated inhibitory effects in hepatocellular carcinoma cells are fortified by genistein.

In a variety of breast cancer cells genistein showed inhibitory effects on their growth at elevated concentration. However, MCF-7 cells, which are estrogen receptor positive, were stimulated by this isoflavone at relatively lower concentrations. Genistein-treated cells manifested cell cycle arrest and apoptosis, which was attributed to Bcl-X$_L$ down-regulation, Bax up-regulation and caspase-3 activation. Synergistic effect was recorded when the defined isoflavone was given in combined mode with fish oil constituent, namely eicosapentaenoic acid (Nakagawa et al. 2000). This outcome signifies that the genistein–EPA combination may culminate in better clinical outcome in case of breast cancer. Genistein along with carmustine (1,3-bis(2-chloroethyl)-1-nitrosourea) was assessed on glioma cells for their growth inhibitory effects. On addition of genistein to carmustine (BCNU), 28–32% and 32–41% increase in colony cytotoxicity and growth inhibition (monolayer) respectively was quantified using the U87 cell line. Further, these effects on the C6 cell line were estimated to be 39–54% and 30–36% (Khoshyomn et al. 2002). From these results it can be inferred that genistein reinforces the effects of carmustine in glioma cells. Another mind-blowing study examined the combined use of genistein and doxorubicin in MCF-7 cells resistant to doxorubicin (MCF-7/Adr cells). The genistein–doxorubicin union demonstrated synergism in these cells. Genistein lowered the resistance of these cells to doxorubicin. Doxorubicins intracellular levels were accumulated on genistein intervention and this effect was irrespective of P-glycoprotein as no influence was shown by this isoflavone on its function. The combination of these two agents substantially triggered cell cycle arrest and apoptotic death in these cells. Strong inhibition was exerted by genistein on HER2/neu at both the protein and transcript level. However, no such effect was noticed on expression of multidrug resistance protein 1 (MDR-1), also termed P-glycoprotein (Xue et al. 2014). The gist taken from these findings is that the genistein–doxorubicin combination exhibits synergistic effect on doxorubicin-resistant cells and this effect is linked to enhanced accumulation of this standard drug in these cells and inhibition of the expression of HER2/neu.

7.2.2 CONVENTIONAL DRUGS AND DAIDZEIN ASSOCIATION

In cancer chemotherapy topotecan is a famous name, but its use is restricted due to off-target effects and resistance developed by cancer cells against it. Daidzein, the natural isoflavone, has been used in conjunction with topotecan and the dual therapy has shown a potent synergistic effect on the tumour cell line. This effect was accompanied by the elevated inhibition of Topo1 by topotecan, the increased number of cells undergoing cell cycle arrest and the eventual apoptosis of a greater number of cells. The daidzein addition reversed the resistance of MCF7/ADR cells to topotecan. The resistance index was lowered from 7.17 to 0.77 and was associated with the inhibition of efflux transporter breast cancer resistance protein (BCRP) and estrogen receptor alpha (Erα), due to which the intracellular levels of topotecan improved. The aforesaid combination also reduced tumour growth not only in the MCF-7 xenograft model but also in MCF7/ADR more efficiently than the topotecan alone (Guo et al. 2019). All this indicates that the anticancer effect of topotecan is greatly improved when it is used jointly with isoflavone daidzein. Daidzein hampered nephrotoxicity induced by cisplatin by mitigating oxidative stress, apoptosis and inflammation and thereby facilitated regeneration of the kidney (Meng et al. 2017). Daidzein induced substantial protection against cisplatin-elicited hepatotoxicity and haemotoxicity by declining a variety of enzymes and by enhancing haemoglobin, RBC, platelets and packed cell volume (Karale & Kamath 2017).

7.3 FLAVANONES IN DOUBLET THERAPY WITH CONVENTIONAL THERAPEUTIC MOLECULES

Naringin, naringenin and hesperidin are the three flavanones that are highly evaluated. They have demonstrated significant activity against various cancer cells (Memariani et al. 2021). For easy understanding, these flavanones will be individually discussed.

7.3.1 NARINGENIN AUGMENTS ANTICANCER EFFECT OF USUAL ANTICANCER DRUGS

Studies have proved that naringenin may serve as a coadjuvant in cervical cancer therapy. The antitumour effect of cisplatin was strengthened by naringenin. A drug combination involving low cisplatin concentrations and naringenin reinforced the effect of cisplatin. The combined regiment markedly reduced the viability of HeLa, intensified the cytotoxic effects and lowered the invasive power of spheroids (Martínez-Rodríguez et al. 2020). Thus, naringenin may serve as a strong adjuvant for curbing the growth and viability of cervical cancer spheroids. Naringenin has manifested puzzling results in altering the function of proteins involved in multidrug resistance. Whether acquired or intrinsic resistance to anticancer agents can be counteracted by this flavanone has been examined. In a variety of cell lines, the impact of naringenin on different aspects of doxorubicin, including its uptake as well as its retention and cytotoxicity, has been scrutinized. Combining naringenin with doxorubicin increased its efficacy. The IC_{50} of combination was 2-fold lower in comparison to IC_{50} quantified when doxorubicin was used as a single agent. In cell lines like MCF-7/DOX and HepG2, naringenin showed no effect in enhancing the sensitivity of these cells to doxorubicin. The intracellular amassing of doxorubicin has been credited to the ability of naringenin to obstruct the activity of multidrug resistance proteins other than P-glycoprotein. Compared to doxorubicin monotherapy the said combination evinced heightened antitumour effect but minimal systemic toxicity under *in vivo* conditions (Zhang et al. 2009). It can be concluded that naringenin improves the anticancer power of doxorubicin by preventing its efflux from cells through specific modulation of multidrug resistance proteins excluding P-glycoprotein.

7.3.2 Naringin Reinforces Effect of Conventional Antineoplastic Agents

The use of doxorubicin against cervical cancer is impeded by its off-target effects, including nephrotoxicity and cardiotoxicity. The improved therapeutic effects of doxorubicin when used in association with naringin have been unbossomed both in cell lines and animal models. The naringin addition to the doxorubicin intervention proved highly effectual in blocking proliferation as compared to the situation when these agents were used singly. Similar results were found under *in vivo* conditions where the combined regime proved to be more effective over monotherapy in restraining HeLa cell tumour growth and apoptosis. The combined treatment mitigated hepatotoxicity, cardiotoxicity and nephrotoxicity in comparison to models treated solely with doxorubicin (Liu et al. 2017). Conclusively, associating naringin with doxorubicin generates enhanced pharmacological effect and significantly soothes toxicity. In nude mice ovarian tumours were generated by inserting ovarian cancer cells. Following this, naringin was tested in this model singly at different concentrations and in concert with cisplatin. Both naringin and expectedly cisplatin reduced tumour growth and this was perceptible from declined tumour weight and growth. The effect of naringin on the depletion of tumour growth followed a concentration-dependent trend. The combined group administered with cisplatin and naringin (2 mg/kg each) also manifested substantial decline in tumour growth. The combined treatment facilitated expression of caspase-3 and 7 and concurrently alleviated Bcl-xL and Bcl-2 expression. Besides, a marked reduction in survivin, cyclin D1 and c-Myc levels was quantified from cells subjected to combined treatment, cisplatin only, or naringin (Cai et al. 2018).

7.3.3 Hesperidin and Hesperetin in Concert with Customary Therapeutics

This citrus flavanone has shown promising antiproliferative effect against multiple cancer cells. The pharmacological effect of hesperidin has been evaluated in cooperation with doxorubicin using HeLa cells as models. Individually, doxorubicin portrayed and IC_{50} of 1000 nM while hesperidin displayed this value as 48 µM. The combination of hesperidin (6 µM) and doxorubicin (500 nM) induced the most potent inhibitory effect against these cells. At concentration of 24 µM, this flavanone induced cell cycle arrest at the G1 phase but its association with doxorubicin (500 nM) resulted in the accumulation of cells in the S phase as well. Both single and combined treatments enhanced Bax expression while Bcl-2 was dampened (Kusharyanti et al. 2011). It is evident from the results that hesperidin has the ability to serve as a co-therapeutic for tackling cervical cancer. Enhanced cytotoxicity was shown by the above-mentioned combination towards breast cancer cells (MCF-7). While individually hesperidin failed to elicit cytotoxicity in these cells, doxorubicin was quite successful in doing so at nanomolar concentration. The highest cytotoxicity was recorded at 100 µM and 200 nM concentration of hesperidin and doxorubicin respectively (Hermawan, Meiyanto & Susidarti 2010). This combined strategy stimulated apoptosis in these cells and these results signify that hesperidin may be used as an adjunct for raising the cytotoxic activity of conventional anticancer agents. Hesperetin, the aglycone of hesperidin, also has anticancer activity and has been used jointly with other chemotherapeutic agents for obtaining additive or synergistic effects (Parhiz et al. 2015). Citrus flavanone hesperetin is multifunctional. It has vasoprotective, antiinflammatory, antioxidant, anticancer and antiallergic activity. This flavanone was tested in collaboration with doxorubicin against murine breast cancer (4T1) cells. Hesperetin exerted cytotoxicity towards these cells and further its combination with conventional doxorubicin demonstrated synergistic cytotoxic effect. This dual intervention enhanced G2/M arrest and subsequent apoptosis (Yunita et al. 2020). Additionally, hesperetin-doxorubicin restrained migration of these cells and this was associated with down-regulation of MMP-9 expression.

7.4 FLAVONOLS IN COMBINATIONAL THERAPY

Like other flavonoids, flavonols have congenital anticancer activity. Flavonols induce death in abnormal cells through a wide range of mechanisms. They have the potential to modulate multiple cellular targets. Isorhamnetin, quercetin, myricetin and kaempferol are certain examples of flavonols that are increasingly being studied with reference to anticancer activity against a myriad of cancers, including gastrointestinal cancer (Ramachandran et al. 2012; Mirazimi et al. 2022). Flavonol-related doubts will become crystal clear by dealing with different flavonol members in a desolated manner.

7.4.1 Isorhamnetin as Co-therapheutic with Endorsed Therapeutics

To improve the treatment of non-small cell lung cancer, it is exigent to try therapeutic approaches that offer better efficacy and tolerability. Isorhamnetin in association with either cisplatin or carboplatin has been checked towards A-549 cells. More strong growth inhibition and apoptosis occurred when isorhamnetin was associated with carbopaltin or cisplatin. Isorhamnetin, either as a single agent or in union with the above-mentioned agents, induced depolymerisation and distortion of microtubules. However, compared to isorhamnetin-treated cells, cells treated using a combined approach spaced more cells in the G2/M phase. The mitochondrial membrane potential was also modulated by the above-mentioned combinations and the stimulation of certain caspases including caspase-3 and 9 was also seen. The migration of A549 cells was considerably abrogated by isorhamnetin–cisplatin and isorhamnetin–carboplatin combinations (Zhang et al. 2015).

7.4.2 Quercetin and Standard Chemotherapeutics in Concert

Cisplatin is considered a frontline therapeutic for oral squamous cell carcinoma. Due to the resistance of these cells towards this therapeutic, new combinations have been tested for circumventing resistance (Najafi et al. 2020). While testing the quercetin–cisplatin combination in the above-defined cells it was noticed that quercetin potentiates the apoptosis induced by cisplatin through NF-κB inhibition and X-linked inhibitor of apoptosis protein (XIAP). Quercetin pre-treatment by way of triggering caspase-8 and 9 improved the intrinsic and extrinsic apoptosis respectively (Li et al. 2019). Thus, using cisplatin–quercetin jointly may result in better clinical outcomes in oral squamous cell carcinoma subjects. The proliferation of nasopharyngeal carcinoma cells was blocked by quercetin and the expression of Ki67 and fatty acid synthase was lowered (Daker, Ahmad & Khoo 2012). Synergistic cytotoxicity was quantified in these cells following quercetin–cisplatin exposure and this was discernible from a combination index below 1. Various drug combinations including quercetin as one of the agents were evaluated on two cervical cancer lines, namely SiHa and HeLa. Of these combinations (quercetin–cisplatin, quercetin–5-FU, quercetin–paclitaxel and quercetin–doxorubicin) only the quercetin–cisplatin combination was found to have stronger antiproliferative effect in comparison to quercetin and cisplatin monotherapy. Also, the impact on cell migration was more intense in quercetin–cisplatin-treated cells. In comparison to cisplatin, the degree of expression of various molecular players such as P-glycoprotein, ezrin, METTL3 and MMP-2 was lower in quercetin–cisplatin-treated cells (Xu et al. 2021). A combination index less than 1 again signified that the cisplatin–quercetin combination results in synergistic effect in these cells. In another study, co-treatment with adriamycin and quercetin offered therapeutic benefit. The effect shown by adriamycin at a high concentration was shown by it at lower concentrations when it was given jointly with quercetin (Shi et al. 2019). Doxorubicin-induced antitumour effects were heightened by quercetin, especially in breast cancer cells, which are highly invasive in character. Besides, the co-treatment hampered the undesirable cytotoxic effects towards typical body cells. While doxorubicin damaged DNA both in cancer and normal body cells quercetin mitigated this damage in the latter cell-type only. The doxorubicin addition sustained quercetin-triggered polynucleation in bellicose tumour cells. The crux is that the doxorubicin–quercetin combination culminates in raised cytotoxic effects towards invasive type breast cancer (Staedler et al. 2011). Further, the toxic effects of doxorubicin towards normal cells are pacified by quercetin when given in combination.

7.4.3 COMBINING CONVENTIONAL DRUG MOLECULES WITH KAEMPFEROL

It is a grim fact that triple-negative breast cancer is not only bellicose but also herculean to treat. Migration in addition to invasion of triple-negative breast cancer (TNBC) cells was strongly suppressed by kaempferol. However, the migration of non-triple-negative breast cancer cells, including MCF-7 and SK-BR-3, was not altered by this flavonol. Intervention with kaempferol lowered the activations of Rac1 and RhoA in these (TNBC) cells. Migration of ER/PR-silence MCF-7 and HER2-silence SK-BR-3 cells and activations of RhoA was also blocked by kaempferol at low dose. Further, kaempferol-induced migration inhibition and modulation of RhoA and Rac1 was reversed on HER2 overexpression in these cells. It was also explored that combining herceptin (inhibitor of HER2) with kaempferol (low dose) strongly obstructed HER2+ breast cancer cell migration (Li et al. 2017). Kaempferol sensitized 5-FU resistant colon cancer cells to 5-FU. The synergistic effect on cell viability was seen when this flavonol was used together with 5-FU. This association increased the apoptotic death in chemosensitive as well as in chemoresistant cells by alleviating ROS generation and modulation of NF-κB, MAPK, JAK/STAT3 and PI3K/AKT (Riahi-Chebbi et al. 2019). Thus, to overcome colon cancer resistance kaempferol may be used along with 5-FU. The dose- and time-dependent inhibitory effect of kaempferol was observed on multiple liver cancer cells. The deeper inhibitory effect on cell viability was achieved when this flavone was used with doxorubicin. The inhibitory effects of co-treatment were also observed on the mitochondrial function, progression of cell cycle. colony formation and DNA damage response (Yang et al. 2021). In addition, this joint therapy blocked migration and invasion of these cells in a highly vigorous manner. This proves that kaempferol has the potential to strongly boost the anticancer activity of standard anticancer drugs.

7.4.4 FISETIN-CONVENTIONAL DRUG COMBINATION

Ovarian cancer cells generate resistance towards cisplatin and, as such, new therapeutics that can be used with this conventional drug are increasingly explored. Further, certain substantial side-effects are associated with cisplatin chemotherapy. Fisetin, like kaempferol and quercetin, belongs to flavonols. Fisetin together with cisplatin has been appraised against the ovarian cancer line resistant to cisplatin (A2780). The proliferation of these cells was robustly inhibited by the fisetin–cisplatin duo. Cells subjected to co-treatment showed chromatin fragmentation (Jafarzadeh, Baharara & Tehranipour 2021). The augmentation of cisplatin cytotoxicity by fisetin has been delineated using human embryonal carcinoma (NT2/D1) cells. Fisetin in combination with low-dose cisplatin increased the cytotoxic effect of the latter. Both these agents singly as well as in combination enhanced the expression of Fas ligand (FasL). The combined approach activated multiple caspases, including caspase-7, 9, 8 and 3. Besides, fisetin–cisplatin as compared to singlet therapy more deeply declined survivin and cyclin B1 and, on the other hand, escalated p21 and p53. In the xenograft model bearing the NT2/D1 tumour, the aforesaid combination was the most effectual in collapsing tumour size (Tripathi et al. 2011). The essence of this study is that the addition of fisetin to cisplatin activates the death receptor as well as the mitochondrial pathway and is thus a propitious tactics to obliterate embryonal carcinoma cells.

7.4.5 STANDARD DRUG MOLECULES ALONG WITH MYRICETIN

Although paclitaxel has been used widely as a remedy for ovarian cancer, these cells often develop resistance by effluxing drug from cells and thus alleviating its intracellular accumulation. The main protein that has cross-talk with paclitaxel resistance is P-glycoprotein encoded by *MDR-1*. Thus, molecules with the potential to inhibit the activity of this protein may potentiate the cytotoxic effects of paclitaxel on ovarian cancer cells. In a couple of ovarian cancer cells, myricetin triggered marked cytotoxicity and apoptosis. This flavonol modulated antiapoptotic and pro-apoptotic proteins. Importantly, pre-treatment of ovarian cancer cells with myricetin enhanced the paclitaxel-invoked cytotoxicity and this effect was accredited to the tendency of myricetin to inhibit MDR-1 (Zheng et al. 2017). Due to a variety of effects, such as antiinflammatory, antioxidant and anticancer, demonstrated by myricetin it has been evaluated towards esophageal cancer (EC9706) cells in conjunction with 5-FU. Cells treated with 5-FU along with myricetin showed reduced proliferation, survival and heightened apoptosis in comparison to cells treated with 5-FU only. This drug combination suppressed the Bcl-2, cyclin D, survivin but elevated p53 and caspase-3 expression. The 5-FU–myricetin combination lowered the growth rate of tumour xenografts (Wang et al. 2014). All this indicates that the myricetin addition enhances the effectiveness of 5-FU towards esophageal cancer cells and thus supports its use in combination with 5-FU for subduing aggressive esophageal malignancies.

7.5 FLAVANOLS IN COOPERATION WITH STANDARD CANCER THERAPEUTICS

Drug paclitaxel is globally used for the treatment of cancer. The use of this anticancer agent is hampered by chemoresistance, thereby emphasizing the desperate need of supporting agents that can overcome the paclitaxel-related resistance (Dan, Raveendran & Baby 2021). The polyphenol (-)-epigallocatechin gallate (EGCG) separated from green tea has been investigated in the above-mentioned context under different conditions (Kim, Quon & Kim 2014). Breast cancer cells were sensitized to conventional paclitaxel by EGCG and this effect was proved to be synergistic. Compared to singlet use of these agents, the combined strategy displayed substantial induction of apoptosis in 4T1 cells. Mice bearing tumours, when subjected to this combination, underwent a sharp reduction in tumour growth, but the results of individual agents were not promising. The glucose-regulated protein 78 (GRP78) comes under the confines of the Hsp70 family and critically regulates endoplasmic reticulum homeostasis (endoplasmic reticulum chaperone) (Ermakova et al. 2006; Li & Lee 2006). Paclitaxel induces this chaperone, which in turn protects breast cancer cells from apoptosis (Li & Lee 2006; Lee 2007). EGCG inhibited the paclitaxel-stimulated induction of GRP78 and eventually strengthened the JNK phosphorylation elicited by paclitaxel even under *in vivo* environment (Luo et al. 2010). This means that EGCG potentiates the anticancer activity of paclitaxel by restraining the expression of GRP78, which is up-regulated following paclitaxel use. Promising results were obtained when the EGCG–doxorubicin combination was examined on A549 cells. Following treatment with this combination, the drug efflux and invasiveness was impaired whereas enhanced drug internalization, stress-triggered damage, cell cycle blockade and the subsequent death of these was recorded (Datta & Sinha 2022). Thus, in lung cancer cells which are non-responsive to doxorubicin, intrinsically coupling herbal flavanol EGCG to doxorubicin reinstates their responsiveness to this standard therapeutic. Due to grievous toxic effects, the platinum-based therapy against colorectal cancer is not exempted from restrictions. Combining herbal molecules with the above-mentioned therapies is tested against various malignancies to come out with combinations that offer maximum therapeutic advantage and concurrently least side-effects. The green tea derivative EGCG has been experimented in concert with oxaliplatin or cisplatin towards HT-29 and DLD-1 cells, keeping in mind whether EGCG addition may pacify the serious off-target effects of these agents. The synergistic antiproliferative effect and apoptosis was estimated with either of these combinations. Oxaliplatin and cisplatin invoked autophagy was boosted by EGCG in both cell lines and this was perceptible from LC3-II protein accumulation, autophagosome formation and the rise in acidic vesicular organelles (Hu et al. 2015). Taken together, EGCG co-treatment with either cisplatin or oxaliplatin induces more than an additive effect in colon cancer cells by deepening their effects on autophagy.

7.6 COMBINING CONVENTIONAL COMPOUNDS WITH ANTHOCYANINS

Herbal molecules are progressively examined as the agents with the potential to raise the therapeutic effect of conventional compounds. As an example, cyanidin-3-O-glucoside (C3G) in association with cisplatin has been scrutinized for pharmacological benefits towards cervical cancer cells. While cisplatin inhibited HeLa cell proliferation by 17.43%, C3G restrained by 34.98% and the combination resulted in 63.38% inhibition. Cisplatin demonstrated an IC_{50} of 18.53 µg/ml, whereas for cisplatin-C3G this value was measured to be 6.435 µg/ml. Although C3G and cisplatin were able to induce arrest at G1 and apoptosis, but the anticancer effects were noticeably higher in case of co-treated cells (Li et al. 2021). These evidences suggest that C3G elevates the tumour-killing activity of cisplatin and this approach may yield fruitful results in addressing the concern of overt toxicities associated with the intervention of platinum-containing compounds. C3G as a single agent and in alliance with 5-FU was administered in mice bearing tumours of lung large-cell carcinoma cells. It was found that C3G either in monotherapy or in union with 5-FU attenuated the growth of tumours, enhanced their apoptosis and alleviated cytokine levels meant for inflammation. The transcript levels of nuclear factor-κB and protein levels of cyclooxygenase-2 were inhibited by this treatment. C3G-5-FU in cooperation exhibited substantial tumour growth inhibition that proved to be synergistic. The combinatorial method manifested a tumour inhibition rate of 80.9%, while 5-FU and C3G showed this rate as 58.4% and 16.7% respectively (Wu et al. 2022). The main crux in these findings is that C3G proves to be highly effectual against lung large-cell carcinoma when it is administered as an adjuvant with well-known anticancer agent 5-FU. Cyanidin 3-glucoside and peonidin 3-glucoside were tested on various cell lines and, among them, HS578T was found to be relatively more sensitive. Both these anthocyanins induced G2/M arrest. Peonidin 3-glucoside reduced the levels of CDK-1, cyclin B1, CDK-2 and cyclin E. The activation of caspase-3 in addition to chromatin compaction and cell death occurred on single agent use of these plant-derived molecules. Combining a low dose of either of these anthocyanins with doxorubicin enhanced its cell growth inhibitory effects besides lowering the TC_{50}. When doxorubicin was used individually TC_{50} was found to be 0.36 µM while in combination with peonidin 3-glucoside or cyanidin 3-glucoside this value was estimated to be 0.19 and 0.08 µM respectively.

The evidence-based discussion on the advantages of using conventional drugs with flavonoids derived from herbal sources has been concluded. As flavonoids fence in flavones, flavanones, flavanols, isoflavones, flavonols and anthocyanins, thus all the main members of these flavonoid subgroups were discussed in combination with standard therapeutics with reference to various preclinical cancer models. The bottom line is that the molecules falling within the confines of flavonoids boost the anticancer activity of conventional medicines. In the majority of cases the effect was found to be synergistic. These herbal molecules not only enhance the cytotoxic effects of certified drugs but also soothe their toxicity towards the usual body cells. This is because when two molecules are used in concert, they induce apoptosis in cancer cells at low doses which are quite tolerable to typical body cells. In certain cases, the addition of a flavonoid member facilitates the normal body cells to revive from toxicity induced by standard cancer therapeutics. The next segment will give a compendium of natural HDACi, excluding non-flavonoids, in collaboration with conventional therapeutics against a broad range of cancers strictly through the prism of certified evidence.

REFERENCES

Alrawi, S. J., Schiff, M., Carroll, R. E., Dayton, M., Gibbs, J. F., Kulavlat, M., Tan, D., Berman, K., Stoler, D. L. & Anderson, G. R. (2006). Aberrant crypt foci. *Anticancer Res 26*: 107–119.

Alshazi, M., Alanazi, W., Alshammari, M., Alotaibi, M., Alhoshani, A., Al-Rejaie, S., Hafez, M. & Shabanah, O. (2017). Neuro-protective effect of rutin against cisplatin-induced neurotoxic rat model. *BMC Complementary and Alternative Medicine 17*: 472.

Aromokeye, R. & Si, H. (2022). Combined curcumin and luteolin synergistically inhibit colon cancer associated with Notch1 and TGF-β signaling pathways in cultured cells and xenograft mice. *Cancers (Basel) 14*: 3001.

Ayyildiz, A., Koc, H., Turkekul, K. & Erdogan, S. (2021). Co-administration of apigenin with doxorubicin enhances anti-migration and antiproliferative effects via PI3K/PTEN/AKT pathway in prostate cancer cells. *Exp Oncol 43*: 125–134.

Bayat Mokhtari, R., Homayouni, T. S., Baluch, N., Morgatskaya, E., Kumar, S., Das, B. & Yeger, H. (2017). Combination therapy in combating cancer. *Oncotarget 8*: 38022–38043.

Cai, L., Wu, H., Tu, C., Wen, X. & Zhou, B. (2018). Naringin inhibits ovarian tumor growth by promoting apoptosis: an in vivo study. *Oncology Letters 16*: 59–64.

Chen, P., Hu, M. D., Deng, X. F. & Li, B. (2013). Genistein reinforces the inhibitory effect of cisplatin on liver cancer recurrence and metastasis after curative hepatectomy. *Asian Pac J Cancer Prev 14*: 759–764.

Cheng, L. & Lai, M. D. (2003). Aberrant crypt foci as microscopic precursors of colorectal cancer. *World J Gastroenterol 9*: 2642–2649.

Choudhury, D., Ganguli, A., Dastidar, D. G., Acharya, B. R., Das, A. & Chakrabarti, G. (2013). Apigenin shows synergistic anticancer activity with curcumin by binding at different sites of tubulin. *Biochimie 95*: 1297–1309.

Daker, M., Ahmad, M. & Khoo, A. S. B. (2012). Quercetin-induced inhibition and synergistic activity with cisplatin – a chemotherapeutic strategy for nasopharyngeal carcinoma cells. *Cancer Cell Int 12*: 34.

Dan, V. M., Raveendran, R. S. & Baby, S. (2021). Resistance to intervention: paclitaxel in breast cancer. *Mini Rev Med Chem 21*: 1237–1268.

Datta, S. & Sinha, D. (2022). Low dose epigallocatechin-3-gallate revives doxorubicin responsiveness by a redox-sensitive pathway in A549 lung adenocarcinoma cells. *J Biochem Mol Toxicol 36*: e22999.

Ermakova, S. P., Kang, B. S., Choi, B. Y., Choi, H. S., Schuster, T. F., Ma, W. Y., Bode, A. M. & Dong, Z. (2006). (-)-Epigallocatechin gallate overcomes resistance to etoposide-induced cell death by targeting the molecular chaperone glucose-regulated protein 78. *Cancer Res 66*: 9260–9269.

Ganai, S. A. (2017). Plant-derived flavone apigenin: the small-molecule with promising activity against therapeutically resistant prostate cancer. *Biomed Pharmacother 85*: 47–56.

Ganai, S. A., Sheikh, F. A. & Baba, Z. A. (2021). Plant flavone chrysin as an emerging histone deacetylase inhibitor for prosperous epigenetic-based anticancer therapy. *Phytotherapy Research 35*: 823–834.

Ganai, S. A., Sheikh, F. A., Baba, Z. A., Mir, M. A., Mantoo, M. A. & Yatoo, M. A. (2021). Anticancer activity of the plant flavonoid luteolin against preclinical models of various cancers and insights on different signalling mechanisms modulated. *Phytother Res 35*: 3509–3532.

Gao, A. M., Zhang, X. Y. & Ke, Z. P. (2017). Apigenin sensitizes BEL-7402/ADM cells to doxorubicin through inhibiting miR-101/Nrf2 pathway. *Oncotarget 8*: 82085–82091.

Goh, Y. X., Jalil, J., Lam, K. W., Husain, K. & Premakumar, C. M. (2022). Genistein: a review on its anti-inflammatory properties. *Front Pharmacol 13*: 820969.

Guo, J., Wang, Q., Zhang, Y., Sun, W., Zhang, S., Li, Y., Wang, J. & Bao, Y. (2019). Functional daidzein enhances the anticancer effect of topotecan and reverses BCRP-mediated drug resistance in breast cancer. *Pharmacol Res 147*: 104387.

Hermawan, A., Meiyanto, E. & Susidarti, R. (2010). Hesperidine increases cytotoxic effect of doxorubicin on MCF-7 cells. *Majalah Farmasi Indonesia 21*: 8–17.

Hu, F., Wei, F., Wang, Y., Wu, B., Fang, Y. & Xiong, B. (2015). EGCG synergizes the therapeutic effect of cisplatin and oxaliplatin through autophagic pathway in human colorectal cancer cells. *J Pharmacol Sci 128*: 27–34.

Huang, S., Yu, M., Shi, N., Zhou, Y., Li, F., Li, X., Huang, X. & Jin, J. (2020). Apigenin and abivertinib, a novel BTK inhibitor synergize to inhibit diffuse large B-cell lymphoma in vivo and vitro. *Journal of Cancer 11*: 2123–2132.

Jafarzadeh, S., Baharara, J. & Tehranipour, M. (2021). Apoptosis induction with combined use of cisplatin and fisetin in cisplatin-resistant ovarian cancer cells (A2780). *Avicenna J Med Biotechnol 13*: 176–182.

Johnson, J. L., Dia, V. P., Wallig, M. & Gonzalez de Mejia, E. (2015). Luteolin and gemcitabine protect against pancreatic cancer in an orthotopic mouse model. *Pancreas 44*: 144–151.

Karale, S. & Kamath, J. V. (2017). Effect of daidzein on cisplatin-induced hematotoxicity and hepatotoxicity in experimental rats. *Indian J Pharmacol 49*: 49–54.

Khan, A. U., Dagur, H. S., Khan, M., Malik, N., Alam, M. & Mushtaque, M. (2021). Therapeutic role of flavonoids and flavones in cancer prevention: Current trends and future perspectives. *European Journal of Medicinal Chemistry Reports 3*: 100010.

Khoshyomn, S., Manske, G. C., Lew, S. M., Wald, S. L. & Penar, P. L. (2000). Synergistic action of genistein and cisplatin on growth inhibition and cytotoxicity of human medulloblastoma cells. *Pediatr Neurosurg 33*: 123–131.

Khoshyomn, S., Nathan, D., Manske, G., Osler, T. & Penar, P. (2002). Synergistic effect of genistein and BCNU on growth inhibition and cytotoxicity of glioblastoma cells. *Journal of Neuro-Oncology 57*: 193–200.

Kim, H. S., Quon, M. J. & Kim, J. A. (2014). New insights into the mechanisms of polyphenols beyond antioxidant properties; lessons from the green tea polyphenol, epigallocatechin 3-gallate. *Redox Biol 2*: 187–195.

Kusharyanti, I., Larasati, L., Susidarti, R. & Meiyanto, E. (2011). Hesperidin increase cytotoxic activity of doxorubicin on HeLa cell line through cell cycle modulation and apoptotis induction. *Indonesian J Cancer Chemoprev* 2: 267–273.

Lee, A. S. (2007). GRP78 induction in cancer: therapeutic and prognostic implications. *Cancer Res 67*: 3496–3499.

Lee, S., Lee, S. K. & Jung, J. (2021). Potentiating activities of chrysin in the therapeutic efficacy of 5-fluorouracil in gastric cancer cells. *Oncology Letters 21*: 24.

Li, J. & Lee, A. S. (2006). Stress induction of GRP78/BiP and its role in cancer. *Curr Mol Med 6*: 45–54.

Li, S., Yan, T., Deng, R., Jiang, X., Xiong, H., Wang, Y., Yu, Q., Wang, X., Chen, C. & Zhu, Y. (2017). Low dose of kaempferol suppresses the migration and invasion of triple-negative breast cancer cells by downregulating the activities of RhoA and Rac1. *OncoTargets and therapy 10*: 4809–4819.

Li, X. & Choi, J.-S. (2007). Effect of genistein on the pharmacokinetics of paclitaxel administered orally or intravenously in rats. *International journal of pharmaceutics 337*: 188–193.

Li, X., Guo, S., Xiong, X. K., Peng, B. Y., Huang, J. M., Chen, M. F., Wang, F. Y. & Wang, J. N. (2019). Combination of quercetin and cisplatin enhances apoptosis in OSCC cells by downregulating xIAP through the NF-κB pathway. *Journal of Cancer 10*: 4509–4521.

Li, X., Huang, J. M., Wang, J. N., Xiong, X. K., Yang, X. F. & Zou, F. (2015). Combination of chrysin and cisplatin promotes the apoptosis of Hep G2 cells by up-regulating p53. *Chem Biol Interact 232*: 12–20.

Li, X., Zhao, J., Yan, T., Mu, J., Lin, Y., Chen, J., Deng, H. & Meng, X. (2021). Cyanidin-3-O-glucoside and cisplatin inhibit proliferation and downregulate the PI3K/AKT/mTOR pathway in cervical cancer cells. *J Food Sci 86*: 2700–2712.

Liu, H., Lee, G., Lee, J. I., Ahn, T. G. & Kim, S. A. (2019). Effects of genistein on anti-tumor activity of cisplatin in human cervical cancer cell lines. *Obstet Gynecol Sci 62*: 322–328.

Liu, X., Yang, X., Chen, F. & Chen, D. (2017). Combined application of doxorubicin and naringin enhances the antitumor efficiency and attenuates the toxicity of doxorubicin in HeLa cervical cancer cells. *Int J Clin Exp Pathol 10*: 7303–7311.

Luo, T., Wang, J., Yin, Y., Hua, H., Jing, J., Sun, X., Li, M., Zhang, Y. & Jiang, Y. (2010). (-)-Epigallocatechin gallate sensitizes breast cancer cells to paclitaxel in a murine model of breast carcinoma. *Breast Cancer Research: BCR 12*: R8.

Martínez-Rodríguez, O. P., González-Torres, A., Álvarez-Salas, L. M., Hernández-Sánchez, H., García-Pérez, B. E., Thompson-Bonilla, M. D. R. & Jaramillo-Flores, M. E. (2020). Effect of naringenin and its combination with cisplatin in cell death, proliferation and invasion of cervical cancer spheroids. *RSC Adv 11*: 129–141.

Memariani, Z., Abbas, S. Q., Ul Hassan, S. S., Ahmadi, A. & Chabra, A. (2021). Naringin and naringenin as anticancer agents and adjuvants in cancer combination therapy: Efficacy and molecular mechanisms of action, a comprehensive narrative review. *Pharmacol Res 171*: 105264.

Meng, H., Fu, G., Shen, J., Shen, K., Xu, Z., Wang, Y., Jin, B. & Pan, H. (2017). Ameliorative effect of daidzein on cisplatin-induced nephrotoxicity in mice via modulation of inflammation, oxidative stress, and cell death. *Oxid Med Cell Longev 2017*: 3140680.

Mirazimi, S. M. A., Dashti, F., Tobeiha, M., Shahini, A., Jafari, R., Khoddami, M., Sheida, A. H., EsnaAshari, P., Aflatoonian, A. H., Elikaii, F., Zakeri, M. S., Hamblin, M. R., Aghajani, M., Bavarsadkarimi, M. & Mirzaei, H. (2022). Application of quercetin in the treatment of gastrointestinal cancers. *Frontiers in Pharmacology 13*.

Najafi, M., Tavakol, S., Zarrabi, A. & Ashrafizadeh, M. (2020). Dual role of quercetin in enhancing the efficacy of cisplatin in chemotherapy and protection against its side effects: a review. *Archives of Physiology and Biochemistry*: 1–15.

Nakagawa, H., Yamamoto, D., Kiyozuka, Y., Tsuta, K., Uemura, Y., Hioki, K., Tsutsui, Y. & Tsubura, A. (2000). Effects of genistein and synergistic action in combination with eicosapentaenoic acid on the growth of breast cancer cell lines. *Journal of Cancer Research and Clinical Oncology 126*: 448–454.

Parhiz, H., Roohbakhsh, A., Soltani, F., Rezaee, R. & Iranshahi, M. (2015). Antioxidant and anti-inflammatory properties of the citrus flavonoids hesperidin and hesperetin: an updated review of their molecular mechanisms and experimental models. *Phytother Res 29*: 323–331.

Prasad, R. & Prasad, S. B. (2021a). Histoprotective effect of rutin against cisplatin-induced toxicities in tumor-bearing mice: Rutin lessens cisplatin-induced toxicities. *Hum Exp Toxicol 40*: 245–258.

Prasad, R. & Prasad, S. B. (2021b). Modulatory effect of rutin on the antitumor activity and genotoxicity of cisplatin in tumor-bearing mice. *Adv Pharm Bull 11*: 746–754.

Qin, T., Zhao, J., Liu, X., Li, L., Zhang, X., Shi, X., Ke, Y., Liu, W., Huo, J., Dong, Y., Shen, Y., Li, Y., He, M., Han, S., Li, L., Pan, C. & Wang, C. (2021). Luteolin combined with low-dose paclitaxel synergistically inhibits epithelial-mesenchymal transition and induces cell apoptosis on esophageal carcinoma in vitro and in vivo. *Phytother Res 35*: 6228–6240.

Ramachandran, L., Manu, K. A., Shanmugam, M. K., Li, F., Siveen, K. S., Vali, S., Kapoor, S., Abbasi, T., Surana, R., Smoot, D. T., Ashktorab, H., Tan, P., Ahn, K. S., Yap, C. W., Kumar, A. P. & Sethi, G. (2012). Isorhamnetin inhibits proliferation and invasion and induces apoptosis through the modulation of peroxisome proliferator-activated receptor γ activation pathway in gastric cancer. *J Biol Chem 287*: 38028–38040.

Riahi-Chebbi, I., Souid, S., Othman, H., Haoues, M., Karoui, H., Morel, A., Srairi-Abid, N., Essafi, M. & Essafi-Benkhadir, K. (2019). The phenolic compound kaempferol overcomes 5-fluorouracil resistance in human resistant LS174 colon cancer cells. *Scientific Reports 9*: 195.

Shi, R., Huang, Q., Zhu, X., Ong, Y.-B., Zhao, B., Lu, J., Ong, C.-N. & Shen, H.-M. (2007). Luteolin sensitizes the anticancer effect of cisplatin via c-Jun NH2-terminal kinase–mediated p53 phosphorylation and stabilization. *Molecular Cancer Therapeutics 6*: 1338–1347.

Shi, Y., Su, X., Cui, H., Yu, L., Du, H. & Han, Y. (2019). Combination of quercetin and adriamycin effectively suppresses the growth of refractory acute leukemia. *Oncology letters 18*: 153–160.

Singh, D., Gupta, M., Sarwat, M. & Siddique, H. R. (2022). Apigenin in cancer prevention and therapy: a systematic review and meta-analysis of animal models. *Crit Rev Oncol Hematol 176*: 103751.

Staedler, D., Idrizi, E., Halamoda, B. & Juillerat-Jeanneret, L. (2011). Drug combinations with quercetin: doxorubicin plus quercetin in human breast cancer cells. *Cancer Chemother Pharmacol 68*: 1161–1172.

Tripathi, R., Samadder, T., Gupta, S., Surolia, A. & Shaha, C. (2011). Anticancer activity of a combination of cisplatin and fisetin in embryonal carcinoma cells and xenograft tumors. *Molecular Cancer Therapeutics 10*: 255–268.

Tsai, K. J., Tsai, H. Y., Tsai, C. C., Chen, T. Y., Hsieh, T. H., Chen, C. L., Mbuyisa, L., Huang, Y. B. & Lin, M. W. (2021). Luteolin inhibits breast cancer stemness and enhances chemosensitivity through the Nrf2-mediated pathway. *Molecules 26*: 6452.

Wang, H., Luo, Y., Qiao, T., Wu, Z. & Huang, Z. (2018). Luteolin sensitizes the antitumor effect of cisplatin in drug-resistant ovarian cancer via induction of apoptosis and inhibition of cell migration and invasion. *J Ovarian Res 11*: 93.

Wang, L., Feng, J., Chen, X., Guo, W., Du, Y., Wang, Y., Zang, W., Zhang, S. & Zhao, G. (2014). Myricetin enhance chemosensitivity of 5-fluorouracil on esophageal carcinoma in vitro and in vivo. *Cancer Cell Int 14*: 71.

Wu, C.-F., Wu, C.-Y., Lin, C.-F., Liu, Y.-W., Lin, T.-C., Liao, H.-J. & Chang, G.-R. (2022). The anticancer effects of cyanidin 3-O-glucoside combined with 5-fluorouracil on lung large-cell carcinoma in nude mice. *Biomedicine & Pharmacotherapy 151*: 113128.

Xu, W., Xie, S., Chen, X., Pan, S., Qian, H. & Zhu, X. (2021). Effects of quercetin on the efficacy of various chemotherapeutic drugs in cervical cancer cells. *Drug Des Devel Ther 15*: 577–588.

Xu, Y., Xin, Y., Diao, Y., Lu, C., Fu, J., Luo, L. & Yin, Z. (2011). Synergistic effects of apigenin and paclitaxel on apoptosis of cancer cells. *PLoS One 6*: e29169.

Xue, J.-P., Wang, G., Zhao, Z.-B., Wang, Q. & Shi, Y. (2014). Synergistic cytotoxic effect of genistein and doxorubicin on drug-resistant human breast cancer MCF-7/Adr cells. *Oncol Rep 32*: 1647–1653.

Yan, V., Wu, T., Leung, S. & To, K. (2020). Flavonoids potentiated anticancer activity of cisplatin in non-small cell lung cancer cells in vitro by inhibiting histone deacetylases. *Life Sci 258*: 118211.

Yang, G., Xing, J., Aikemu, B., Sun, J. & Zheng, M. (2021). Kaempferol exhibits a synergistic effect with doxorubicin to inhibit proliferation, migration, and invasion of liver cancer. *Oncol Rep 45*: 32.

Yosefi, S., Pakdel, A., Sameni, H., Semnani, V. & Bandegi, A. (2022). Chrysin-enhanced cytotoxicity of 5-fluorouracil-based chemotherapy for colorectal cancer in mice: investigating its effects on cyclooxygenase-2 expression. *Brazilian Journal of Pharmaceutical Sciences 58*: 1763.

Yunita, E., Muflikhasari, H. A., Ilmawati, G. P. N., Meiyanto, E. & Hermawan, A. (2020). Hesperetin alleviates doxorubicin-induced migration in 4T1 breast cancer cells. *Future Journal of Pharmaceutical Sciences 6*: 23.

Zhang, B. Y., Wang, Y. M., Gong, H., Zhao, H., Lv, X. Y., Yuan, G. H. & Han, S. R. (2015). Isorhamnetin flavonoid synergistically enhances the anticancer activity and apoptosis induction by cis-platin and carboplatin in non-small cell lung carcinoma (NSCLC). *Int J Clin Exp Pathol 8*: 25–37.

Zhang, F., Du, G., Zhang, L., Zhang, F., Lu, W. & Liang, W. (2009). Naringenin enhances the anti-tumor effect of doxorubicin through selectively inhibiting the activity of multidrug resistance-associated proteins but not P-glycoprotein. *Pharmaceutical Research 26*: 914–925.

Zheng, A. W., Chen, Y. Q., Zhao, L. Q. & Feng, J. G. (2017). Myricetin induces apoptosis and enhances chemosensitivity in ovarian cancer cells. *Oncology Letters 13*: 4974–4978.

8 Standard Drugs in Cooperation with Natural Non-Flavonoid Histone Deacetylase Inhibitors for Stunning Therapy Against Aggressive Malignancies

Non-flavonoid natural histone deacetylase inhibitors (HDACi) like flavonoid HDACi have demonstrated anticancer activity as single agents as well as in collaboration with standard drugs such as paclitaxel, docetaxel, 5-Fluorouracil, doxorubicin, gemcitabine, cisplatin and others. However, as monotherapeutic agents these non-flavonoids exert only mild to moderate cytotoxicity against malignant cells. Coupling these agents with conventional drug molecules not only potentiates the anticancer activity of both agents, but also safeguards the usual cells from toxic effects. Certain combinations prove effective when given simultaneously, while other show enhanced benefit only through sequential administration. Although the majority of combinations show profound synergistic or additive effect, few combinations may culminate in antagonistic effect. The latter combinations are not entertained for therapeutic purpose but are crucial to rule out co-treatments with futile outcome.

Among natural histone deacetylase inhibitors (HDACi) having non-flavonoid architecture organosulfur compounds, stilbenes, bromotyrosines, short chain fatty acids and isothiocyanates are prominent. Diallyl trisulfide (DATS), bis (4-hydroxybenzyl) sulfide, S-allyl mercaptocysteine (SAMC) and diallyl disulfide (DADS) come within the enclosure of organosulfur compounds (Ruhee et al. 2020). Resveratrol as well as piceatannol are two stilbenes possessing histone deacetylase (HDAC) inhibitory activity (Venturelli et al. 2013; Kukreja & Wadhwa 2014; Hsu et al. 2016). While psammaplins fall within the boundary of bromotyrosine derivatives, butyrate, propionate and valeric acid are typical examples of natural short chain fatty acids (Kim et al.

DOI: 10.1201/9781003294863-8

1999; Ohira, Tsutsui & Fujioka 2017; Silva, Bernardi & Frozza 2020). Coumarins are vast and encompass esculetin, novobiocin, esculin, dicumarol and many more (Venugopala, Rashmi & Odhav 2013; Küpeli Akkol et al. 2020).

8.1 COMBINATORIAL THERAPEUTIC APPROACH INVOLVING ORGANOSULFURS

Garlic-derived organosulfide, namely DATS, has strong chemopreventive effect. Its combination with cisplatin has been assessed towards gastric cancer (BGC-823) cell tumours. DATS reduced the viability of these cells through modulation of cyclins and caspases and Bcl-2. The administration of DATS in cooperation with cisplatin in xenograft mice bearing the BGC-823 tumour showed increased activity against the tumour as compared to the group subjected to DATS only. Fewer non-desirable effects were noticed in mice that were given co-treatment. Gastric cancer cells exposed to DATS evinced p38 and JNK/MAPK activation besides impairment in the Nrf2/Akt pathway (Jiang et al. 2017). From these proofs it is clear that DATS has the ability to serve as adjuvant with cisplatin for effective tackling of gastric malignancies. While experimenting the effect of DATS and diallyl disulfide (DADS) on sensitizing the triple-negative breast cancer cells resistant to paclitaxel (TNBC/PR) the verdict was that DATS and DADS do not induce cytotoxicity in the epithelial cell line (non-tumorigenic) but exert pronounced toxicity towards MDA-MB 468 PR and MDA-MB 231 PR cells. Both these garlic derivatives by way of enhancing ROS levels facilitated stimulation of caspase-3 as well as caspase-9 (Marni et al. 2022).

Skin cancer cells are associated with a noticeable decline in the expression of p53, which in turn modulates cell cycle arrest and thus facilitates proliferation of these cells. When skin cancer cells were given single agent treatment with diallyl sulfide (DAS) or paclitaxel, or a combination of both, a noticeable increase in levels of p53 was quantified. The combined strategy proved more effective over single agent intervention in modulating the p53 pathway and eventual decline in viability of these abnormal cells (Muninathan 2021). It was notified that aged garlic extract sensitizes human breast cancer cells to doxorubicin. While no substantial sensitizing effect was observed at 10 mg/mL of this extract, noticeable enhancement in doxorubicin activity was recorded at 50 and higher (93 mg/mL) concentration. Doxorubicin in singlet therapy demonstrated an inhibitory concentration 50 (IC_{50}) of 1.85 μM, which was reduced to 0.962 and 0.999 μM when this standard drug was combined with the aforementioned effective concentrations respectively. Enhanced apoptotic activity on co-treatment has been ascribed to the inhibition of doxorubicin effluxing activity of P-glycoprotein and augmentation of doxorubicin uptake by these cells (Alkreathy et al. 2020). Both nuclear factor-kappaB (NF-κB) and metallothionein 2A (MT2A) have considerable links with cancer chemosensitivity as well as cancer onset. Depleted levels of IκB-α and metallothionein 2A were measured in gastric cancer and these underexpressed levels corresponded with poor prognosis in patients of gastric cancer. Experiments were done on multiple cell lines of gastric cancers, including AGS, SGC7901 and BGC823. Proliferation of these lines was substantially blocked by DATs and this antiproliferative effect was proved through colony formation assay where saline-treated cells gave rise to a greater number of colonies over DATs-treated gastric cancer cells. Further, this garlic-separated molecule was evaluated on the mice model bearing the subcutaneous BGC823 xenograft and following its administration intraperitonally, the volume and weight of the tumour substantially shrank in comparison to the control mice. However, no reduction in the weight of tumour-carrying mice was observed after DATs injection, clearly indicating its clinical relevance. Bath application of DATs elevated IκB-α (total), which in turn depleted cyclin D1, the target gene of NF-κB. The expression of XIAP and phosphorylation of p-65 were reduced and, on the other side, the proapoptotic Bax expression displayed an antiparallel trend to that of XIAP. This treatment was also accompanied with selective induction of metallothionein 2A in gastric cancer cells and no alteration was seen in the expression of metallothionein 1 isoforms. Deeper studies revealed that DATs regulate the metallothionein 2A by following the epigenetic route. Gastric cancer cells treated with DATs showed reduced protein levels of HDAC1 as well as HDAC2 and enhanced acetylation status of H4K5 (histone H4 lysine 5) and H3K9 (histone H3 lysine). Importantly, DATs in company with doxorubicin synergistically reduced the growth of gastric cancer cells. Cell cycle arrest induced by singlet treatment of DATs was further heightened when doxorubicin was added to this garlic-derived molecule. Compared to individual agents, the defined combination enhanced the apoptosis in the above-mentioned cells (Pan et al. 2016).

8.2 STILBENES AS ADJUVANT THERAPEUTICS WITH CONVENTIONAL ANTINEOPLASTIC AGENTS

The profound cytotoxic effects of stilbenes have been proved on various cancer models. These non-flavonoid natural HDACi as single agents have proved effective to a certain degree, but it has been found that when they are used as adjuvants with certain standard chemotherapeutics they raise their cytotoxicity towards cancer cells (Sirerol et al. 2016; Lee et al. 2019).

8.2.1 RESVERATROL-CONVENTIONAL DRUG COMBINATION

Breast cancer cells that are triple-negative and resistant to paclitaxel were found to get sensitized by coupling this anticancer agent with the herbal agent resveratrol. The MDA-MB-231/PacR) cell line generated from MDA-MB-231 has a 12-fold higher resistance for paclitaxel. It was found that resveratrol inhibited the proliferation of these lines and induced apoptosis. Paclitaxel-induced cytotoxic effects were heightened by this stilbene in both cell lines. Moreover, the main reason of paclitaxel resistance was accredited to be the enhanced expression of MDR1 and CYP2C8 (Sprouse & Herbert 2014). This indicates that resveratrol as a standalone and together with approved therapeutics may prove as effective even against triple-negative breast cancer cells resistant to paclitaxel. A cation channel non-selective but permeable to calcium, namely transient receptor potential melastatin 2 (TRPM2), has a critical role in a variety of physiological processes including apoptosis (Hecquet et al. 2014; Cao et al. 2015). This channel being polymodal can be actuated through a broad spectrum of stimuli (Huang et al. 2018). The combination of resveratrol with paclitaxel reduced the viability of glioblastoma (DBTRG) cells and, for this effect, stimulation of TRPM2 channel by this combination was found to be critical (Öztürk et al. 2019). The effects of this combination were also examined on human liver cancer cells (Hepg2). Resveratrol at concentration of 10 μg/mL showed growth inhibitory effect neither on normal cells nor on HepG2 cells, while paclitaxel 5 or 10 μg/mL displayed similar results in normal cells. However, paclitaxel (5 μg/mL) inhibited growth of HepG2 cells, which was further potentiated when this drug was given in concert with resveratrol. Maximum apoptosis was recorded when resveratrol and paclitaxel were used at concentrations of 5 μg/mL and 10 μg/mL respectively (Jiang et al. 2017). From the outcome it can be deduced that the paclitaxel-stimulated anticancer effects can be intensified by its alliance with resveratrol in case of liver cancer. In other terms this stilbene has the ability to sensitize liver cancer cells to paclitaxel intervention.

One more research group investigated the benefits of this combination in the non-small cell lung cancer (A549) cell line. It was confirmed that 10 μg/ml resveratrol have no impact on these cancer cells and on normal MRC-5 cells. Paclitaxel also proved neutral towards normal cells at both concentrations (5 μg/ml and 10 μg/ml). Both these concentrations were able to restrain the growth of A549 cells and this growth inhibition was further augmented by incorporation of resveratrol. Paclitaxel together with resveratrol enhanced the expression of COX-2 not only at the transcript level but also at the protein level. Further, the modulation of genes related to COX-2, such as MMP-1, VEGF, MMP-9, p21, p53, TIMP1, TIMP2 and many more, was quite noticeable in cells subjected to dual treatment (Kong et al. 2017). The meaning that can be extracted from these results is that resveratrol has the tendency to function as an anticancer boosting agent for paclitaxel. Doxorubicin comes under the frequently used chemotherapeutics employed for circumventing breast cancer, but its efficacy is attenuated by acquired resistance generated by these cells. This often results in poor prognosis and the reappearance of disease (Lovitt, Shelper & Avery 2018). The impact of resveratrol on breast cancer cells resistant to doxorubicin was checked apart from its influence on epithelial to mesenchymal transitions of breast cancer cells which are resistant to adriamycin (MCF7/ADR). Associating the above-mentioned stilbene with doxorubicin not only obstructed cell growth but also inhibited migration and facilitated apoptotic death. The expression of SIRT1 was elevated by resveratrol, which in turn resulted in β-catenin destruction (Jin et al. 2019). By following this mechanism, the epithelial to mesenchymal transition characteristics of adriamycin-resistant cells were reversed by resveratrol. The gastric cancer cell line resistant to doxorubicin (SGC7901/DOX) was given combined treatment with resveratrol–doxorubicin. Although these cells have developed drug-resistant and enhanced migratory behaviour, intervention with resveratrol reinstated their sensitivity to doxorubicin. While these cells manifested no change in survival following doxorubicin exposure, resveratrol had considerable effect on the viability of these cells. The synergistic effect on cell survival inhibition was profound when the combinatorial approach involving this herbal agent and doxorubicin was used on cells. The single agent use of doxorubicin facilitated the migration of SGC7901/DOX cells by 25.9%, while a 53.13% reduction in migration of these cells was observed after treatment with resveratrol. Most importantly, stronger inhibition on cell migration was estimated when the resveratrol and doxorubicin combination was opted.

Singlet therapy with doxorubicin increased the expression of β-catenin and vimentin but mitigated E-cadherin. The expression of E-cadherin and down-regulation of vimentin was induced by resveratrol under both situations whether used singly or jointly with doxorubicin. However, no substantial expression changes were seen in β-catenin while resveratrol was used alone. When resveratrol was coupled with doxorubicin, it inhibited the doxorubicin-elicited elevation of this cadherin protein complex subunit (β-catenin). Besides, the combined strategy markedly declined the vimentin expression. Resveratrol-doxorubicin was also evaluated on SGC7901/DOX xenografts carrying mice and it was found that this combination reduced tumour growth synergistically. While a 50.44% reduction in tumour dimensions was measured in mice treated with doxorubicin only, 58.42% inhibition was observed in mice administered with resveratrol. Notably, the combined tactics depleted the tumour volume by 86.97%, clearly signifying synergistic effect. This elevated therapeutic effect of combination has been attributed to the activation of phosphatase and tensin homolog (PTEN) (Xu et al. 2017). Thus DOX-triggered EMT of these cells is inhibited by resveratrol through E-cadherin up-regulation and by depleting the levels of vimentin as well as β-catenin. The combinatorial approach involving resveratrol and 5-FU was tested on the colorectal cancer line and it was noticed that this approach raises apoptosis in these cells in comparison to monotherapy involving resveratrol. Epithelial to mesenchymal transition was also shown by the aforesaid approach. The duo alleviated the phosphorylated-STAT3 and inhibited its attachment to human telomerase reverse transcriptase promoter (Chung et al. 2018). This specifies that the plant stilbene resveratrol augments the pro-apoptotic and antitelomeric effects of 5-FU resulting in resensitization of colorectal cancer cells to the latter. This combination was also probed for its beneficial effects in skin cancer lines and an animal model. Resveratrol potentiates the effect of 5-FU by way of invoking cell cycle arrest at the S-phase. The defined combination was found to have synergistic effect and this association resulted in the regression of the tumour in the mouse model of skin cancer. Dual treatment was more effective under both conditions (*in vitro* and in animal model). Together, resveratrol and 5-FU raised the percentage of apoptotic cells, caused activation of caspase-3, decreased the ratio of Bax to Bcl-2 (Dun et al. 2015). These observations suggest that coupling resveratrol to 5-FU results in augmented therapeutic benefit in colorectal cancer cells and the same might be true for other cancer types. It has been confirmed that resveratrol induces apoptosis even in those colon cancer cells that do not contain p53. Colon cancer cells when exposed to 5-FU underwent apoptosis, which was substantially repealed in the absence of p53. The above-mentioned combination displayed synergistic anticancer effect on these cells irrespective of the status of p53 (Chan et al. 2014). The therapeutic effects of 5-FU singly as well as in cooperation with resveratrol were explored using colon cancer cells SW480 and LoVo as models. The survival rate of colon cancer cells subjected to 5-FU was found to be lesser. This effect was more noticeable when 5-FU was used in collaboration with resveratrol (Huang et al. 2019). Moreover, while probing the effect of 5-FU and its combination with resveratrol on induced colon cancer, it was found that resveratrol was better over 5-FU in terms of therapeutic effect against this cancer in rats. The cytotoxic effect induced by 5-FU was depleted by the combination, but the typical appearance of colon tissue was elevated (Soliman, Farrag & Khaled 2018).

For multiple cancers 5-FU serves as chemotherapeutic, but in clinical settings its use is impeded by resistance mechanisms developed by cells. The antiangiogenic effects of this drug were evaluated alone and in conjunction with resveratrol in the melanoma model (B16 murine). The combined treatment proved more effective over either of the drugs in obstructing cell proliferation and this antiproliferative effect accorded with modulation of AMP-activated protein kinase (AMPK), vascular endothelial growth factor (VEGF), vasodilator-stimulated phosphoprotein (VASP) and cyclooxygenase-2. Tumour growth was collapsed by this combined method and this coincided with alteration in the degree of expression of VASP, AMPK and VEGF. In comparison to the control group, the combined intervention decreased the count of microvascular vessels (Lee et al. 2015). Taken together, the combined approach suppressed not only the growth of cells but also induced the antiangiogenic effect in the murine model bearing the tumour composed of B16 cells.

Multiple studies have confirmed that the apoptosis induced by gemcitabine is augmented by resveratrol in pancreatic cancer cells (Jiang et al. 2016). The combination of the two has been investigated on the Panc-1 and MiaPaCa-2 cell lines. While these agents individually inhibited the viability of these cells, more substantial effect was quantified in cells co-treated with the above-mentioned agents. Compared to gemcitabine- or resveratrol-treated cells, a considerable lesser number of colonies was confirmed from the co-treated group. The marked elevation in apoptosis was noticeable in co-treated cells, which signifies that resveratrol strengthens the gemcitabine-elicited apoptosis. While Bax expression was escalated, PCNA levels were lowered by resveratrol–gemcitabine. The sterol regulatory element binding protein 1 (SREBP1) has a critical role in stemness induced by gemicitabine. Pancreatic cancer cells got sensitized to gemcitabine by resveratrol-induced inhibition of SREBP1 (Zhou et al. 2019). These observations tempt the conclusion that gemcitabine-invoked stemness is antagonized by resveratrol and this effect is mediated by SERBP1 suppression. A protein belonging to the outer mitochondrial membrane, namely nutrient-deprivation autophagy factor-1 (NAF-1), is crucial for the metabolism of calcium as well as for inhibiting autophagy and apoptosis. Experimental evidence endorses the theory that resveratrol enhances the sensitivity of pancreatic cancer cells to gemcitabine. Mia paca-2 and Panc-1 cells co-treated with these agents markedly facilitated the apoptosis of these cells. Furthermore, it was delineated that gemcitabine accompanied with resveratrol was comparatively more effective in inhibiting the proliferation of these cells in comparison to alone treatments. Notably, the experiments proved that resveratrol-induced sensitivity to gemcitabine is arbitrated through the inhibition of NAF-1. Resveratrol induced ROS generation, which in turn stimulated nuclear factor erythroid 2–related factor 2 (Nrf2) signalling and subsequent NAF-1 inhibition (Cheng et al. 2018). An interesting study was performed on PaCa cells and nude mice bearing the PaCa tumour. In this study gemcitabine and resveratrol were used either in a desolated condition or together. A significant increase in apoptosis was seen under both situations in comparison to isolated treatments. Gemcitabine as monotherapeutic caused a sizeable alleviation in the expression of VEGF-B over the control. Pharmacological intervention with the duo substantially mitigated the VEGF-B expression compared with resveratrol as lone agent. The overexpression of VEGF-B in these cells noticeably attenuated the anticancer effect of resveratrol–gemcitabine (Yang et al. 2021). In a nutshell, the enhanced apoptotic effect induced by the gemcitabine–resveratrol combination is partly interceded by VEGF-B down-regulation. To subdue drug resistance in ovarian cancer, a combination of resveratrol with either oxaliplatin or cisplatin has been given a try. The synergistic effect demonstrated by these combinations proved to be schedule dependent. The maximum synergism was achieved when resveratrol was first administered and, after the elapse of two hours, cisplatin or oxaliplatin were administered. However, when resveratrol and either of the two platinum compounds were administered in the form of bolus the least synergism was attained (Nessa et al. 2012). All this indicates that resveratrol pretreatment induces sensitivity in ovarian cancer cells for apoptosis invoked by cisplatin or oxaliplatin.

8.2.2 PICEATANNOL TOGETHER WITH CERTAIN CONVENTIONAL DRUGS

The deregulation of signals meant for survival and tumour suppressors plays a critical role in the resistance of ovarian cancer to cisplatin. Piceatannol, the metabolite of the famous stilbene resveratrol, has shown promising outcomes against distinct cancers, especially in combination with approved therapeutics (Potter et al. 2002; Kershaw & Kim 2017). This resveratrol metabolite elevated the sensitivity of ovarian cancer cells to cisplatin by modulating certain premier players involved in chemoresistance. NOXA, the BH3-only protein, plays a central role in controlling decisions related to cell death and is considered as the premier p53 response gene (Ploner, Kofler & Villunger 2008). The expression of NOXA, a proapoptotic protein, is promoted by this stilbene besides facilitating the degradation of the X-linked inhibitor of the apoptosis protein (XIAP) and stimulation of caspase-3. Dynamin-related protein 1 (*DRP1*) is known to facilitate mitochondrial fission (Favaro et al. 2019). Piceatannol induced DRP1-backed mitochondrial fission, which in turn ended up with highly effectual apoptosis induction. The greater decline in tumour weight was measured in *in vivo* tumour models, which were administered with both these agents as compared to those given either resveratrol or cisplatin (Farrand et al. 2013). Thus, piceatannol heightens cisplatin sensitivity in ovarian cancer models by modulating XIAP, p53 and last, but no way least, mitochondrial fission. While exploring the mechanism involved in augmentation of gemcitabine efficacy by piceatannol against NSCLCs, it was perceptible that resveratrol lowers the IC_{50} of gemcitabine towards these cells. Bad and Bak, the two proapoptotic players, manifested increased expression on gemcitabine use and prior exposure of these cells to piceatannol further raised the Bak expression (Xu & Tao 2014). It can be verdicted that the pre-treatment of NSCLC cells with stilbene piceatannol intensified the cytotoxic effects of gemcitabine through up-regulation of Bak.

An interesting study has been done where piceatannol has been used alone and in concert with cisplatin. This study included various chemosensitive cell lines and their corresponding chemoresistant variants. In some of these lines p53 was a wild type, but in some its status was mutated. Piceatannol as a single agent depleted the viability of chemosensitive lines and potentiated the cytotoxicity of cisplatin in these cells. This stilbene also sensitized the chemoresistant version of the OV2008 line possessing canonical p53 to the above-mentioned platinum compound. Further, significant alleviation of cell viability was apparent in OVCAR-432 cells which, although chemosensitive, possess mutated p53. The combination index was estimated using OV2008 as a representative cell line and it was found that apoptosis induced by piceatannol–cisplatin was found to be additive as signified by the value of 1.06. The beneficial effects of co-treatment were associated with up-regulation of NOXA in addition to stabilization and activation of p53. When tested on tumour growth under *in vivo* conditions these agents showed no effect when used singly. But administration of piceatannol together with cisplatin demonstrated a noticeable collapse of tumour growth. The tumour weight of mice subjected to co-treatment was substantially reduced. The typical tubular morphology of mitochondria was transformed into a fragmented form to a greater degree or wholly in some tumours recovered from mice that were given the combined treatment of piceatannol and cisplatin (Farrand et al. 2013). From these results it can be deduced that the elevated therapeutic benefits obtained by the piceatannol–cisplatin combination are achieved at the molecular level through induction of NOXA, p53 activation as its increased stability due to down-regulation of MDM2 (mouse double minute 2).

8.3 CONVENTIONAL DRUGS IN UNION WITH SHORT CHAIN FATTY ACIDS

Produced from indigestible fibre through the activity of gut flora, short chain fatty acids are structurally less complex than other HDACi. GPR41 and GPR43 are the examples of G-protein coupled receptors. Short chain fatty acids serve as ligands for these receptors, and these receptors are critical for cell proliferation modulation and apoptosis induction. Keeping this in view, it was examined whether short chain fatty acids, especially propionate, increases the cytotoxic effects of cisplatin through activation of either GPR41 or GPR43 in hepatocellular carcinoma cells. Propionate increased the cisplatin-induced apoptosis in these cells in comparison to cisplatin alone. It was found that combinatorial treatment elevated the levels of acetylated H3 and active caspase-3 and this effect was found to be reliant on GPR41. Further, the expression of various HDACs (HDAC3-6 and HDAC8) underwent down-regulation, while TNF-α demonstrated substantial up-regulation. The propionate–cisplatin combination synergistically induced apoptosis in HepG2 cells by heightening the TNF-α expression through depletion of HDAC levels by way of GPR41 signaling (Kobayashi et al. 2018). It is obvious from the results that the GPR41-specific agonist may yield better outcomes in combination with standard therapeutics of hepatocellular carcinoma.

8.4 COUMARINS AS ADJUVANTS FOR STANDARD ANTINEOPLASTIC AGENTS

Like other non-flavonoids, coumarins have also evinced enhanced preclinical outcomes with other therapeutics employed in cancer chemotherapy. Novobiocin, an amino-coumarin, augmented the activity of cisplatin against human small cell lung carcinoma cells. Two cell lines, namely GLC4 and another one resistant to cisplatin GLC4/CDDP, were selected for the study. While constant incubation with this coumarin augmented the cisplatin-induced cytotoxicity in GLC4 by 1.9 fold, no effect was noticed on GLC4/CDDP cells. On the other hand, brief incubation with amino-coumarin novobiocin escalated the cytotoxic effects of cisplatin, but the effects were more profound in GLC4 as compared to GLC4/CDDP. Novobiocin increased the cisplatin-elicited cytotoxicity in GLC4/CDDP only by a factor of 2.8 and in GLC4 by 4.1. This differential effect in the latter approach was possibly due to the increased count of cisplatin-triggered DNA interstrand cross-links in GLC4 cells over GLC4/CDDP. The limited augmentation of cisplatin cytotoxicity in resistant cells has been ascribed to their previously built higher DNA topoisomerase (Topo II) activity. This is quite logical as novobiocin hampers the catalytic activity of this Topo (de Jong et al. 1993).

Alkylating agents cause DNA double strand cross-links finally resulting in the cytotoxicity of cancer cells. Novobiocin, which is known to inhibit the catalytic function of DNA topoisomerase II, boosts the cytotoxicity of these agents. Novobiocin was used in concerted fashion with carmustine or cisplatin and it was observed that these combinations exerted synergistic cytotoxic effects in Chinese hamster ovary cells. As compared to cisplatin as a single agent, novobiocin–cisplatin-treated cells showed a six-fold increase in interstrand cross-links. Further, this effect was found to be sequence-dependent. The said combinations markedly decreased the tumour growth of murine fibrosarcoma and this was not associated with elevated toxicity to the tumour-bearing host (Eder et al. 1987). Thus, associating novobiocin with different alkylating agents may prove promising in subduing tumour growth and miti-gating toxicity towards normal body cells. Another interesting study revealed that the combination of novobiocin with lonidamine or topotecan increases the melphalan-induced cytotoxicity (Schwartz et al. 1993). Human breast cancer cells with acquired resistance against topotecan and mitoxantrone (MCF7/TPT300) were treated with novobiocin for enhanced cytoxicity. The inference was that novobiocin augments the cytotoxic effects of both mitoxantrone and topotecan towards these cells and that too at concentrations that are defined as clinically relevant. The intracellular amassing of topotecan in MCF7/TPT300 cells was raised by this amino-coumarin and this was due to the inhibition of efflux of topotecan from cells by novobiocin (Yang, Chen & Kuo 2003). Breast cancer-resistance protein (BCRP) comes under the ATP-binding cassette transporters and being an efflux transporter makes breast cancer cells resistant to anticancer agents (Mao & Unadkat 2015). In a pair of cell lines including one breast cancer line novobiocin obstructed topotecan efflux due to BCRP by inhibiting the function of this protein. An important study investigated the impact of novobiocin in sensitizing drug-resistant and drug-sensitive melanoma (B16) cells to vinblastine and colchicine. The COL/R cell line was chosen for colchicine resistance and this cell line was confirmed to have the phenotype of the multidrug resistant type. Additionally, these cells were also resistant to adriamycin and vinblastine (cross-resistance) and were found to have P-glycoprotein overexpressed. A synergistic antiproliferative effect was seen in parental B16 cells when novobiocin–colchicine or novobiocin–vinblastine combinations were employed. The exposure of COL/R cells to novobiocin attenuated their resistance to mitotic poison colchicine (Nordenberg et al. 1994). This experimental evidence indicates that novobiocin sensitizes not only drug-sensitive cells to micro-tubule depolymerizing agents but also melanoma cells exhibiting multidrug resistance.

Multidrug resistance, like other cancers, also occurs in patients of acute myeloid leukemia. Currently, natural products are used in combination with approved therapeutics for achieving maximum therapeutic advantage. Coumarin, a plant-derived molecule, along with doxorubicin as a novel combination has been assessed for circumventing drug resistance in case of haematological malignancy, namely acute myeloid leukaemia. In HL60 line, monotherapy involving coumarin induced marked death, but the same was not true for the drug-resistant version of this line (HL60/ADR). The combination of the two induced apoptosis even in acute myeloid leukaemia cells that were not drug sensitive. Substantial apoptosis induced by the dual treatment was accompanied with enhanced p53, caspase-3 stimulation, PARP cleavage and Bcl-2 inhibition (Al-Abbas & Shaer 2021). Tackling cancer with low-dose therapeutics is achievable and induces feeble toxicity towards typical body cells. Taxol, the known anticancer molecule, is used jointly with other natural molecules for attaining better pharmacological effect and for dampening toxicity. Esculetin, a simple coumarin, was given in union with taxol to human hepatoma cells (HepG2). As a sole agent, esculetin exhibited antiproliferative effect that was associated with the suppression of ERK activation as well as p38 MAPK. Co-treatment with both taxol and esculetin exerted relatively more apoptotic effect than obtained with taxol as a single agent. The combined treatment escalated the Bax expression, cytosolic cytochrome C level, Fas ligand, Fas and activation of caspase-8 and 3. Further it was delineated that the heightening of taxol-induced apoptosis due to this coumarin has cross-talk with ERK signalling (Kuo et al. 2006).

The escalated therapeutic advantages of non-flavonoid but natural HDACi have been described deeply. In the beginning, the organosulfur compounds, such as DADS, DATS and others, were elaborated in the context of the combined therapeutic approach with standard drugs often used for tackling monotonous cancers. This was followed by stilbenes like resveratrol, and its metabolite, namely piceatannol. The benefits of natural short chain fatty acid HDACi, especially in collaboration with conventional therapeutics, towards various cancer types were illuminated. Last, but not least, coumarins encompassing novobiocin and esculetin were discussed thoroughly as adjuvants with standard chemotherapeutics. In other words, non-flavonoid HDACi irrespective of the group to which they belong boosted the anticancer potential of standard molecules, attenuated their toxicity towards usual body cells and raised their intracellular levels in atypical cancer cells.

REFERENCES

Al-Abbas, N. S. & Shaer, N. A. (2021). Combination of coumarin and doxorubicin induces drug-resistant acute myeloid leukemia cell death. *Heliyon 7*: e06255.
Alkreathy, H., AlShehri, N., Kamel, F., Alghamdi, A., Esmat, A. & Karim, S. (2020). Aged garlic extract potentiates doxorubicin cytotoxicity in human breast cancer cells. *Tropical Journal of Pharmaceutical Research 19*: 1669–1676.
Cao, Q. F., Qian, S. B., Wang, N., Zhang, L., Wang, W. M. & Shen, H. B. (2015). TRPM2 mediates histone deacetylase inhibition-induced apoptosis in bladder cancer cells. *Cancer Biother Radiopharm 30*: 87–93.
Chan, J., Phoo, R., Pervaiz, S. & Lee, S. C. (2014). Combination effects of resveratrol and 5-fluorouracil on human colon cancer cells. *Clinical Cancer Research 13*: A14-A14.

Cheng, L., Yan, B., Chen, K., Jiang, Z., Zhou, C., Cao, J., Qian, W., Li, J., Sun, L., Ma, J., Ma, Q. & Sha, H. (2018). Resveratrol-induced downregulation of NAF-1 enhances the sensitivity of pancreatic cancer cells to gemcitabine via the ROS/Nrf2 signaling pathways. *Oxid Med Cell Longev 2018*: 9482018.

Chung, S. S., Dutta, P., Austin, D., Wang, P., Awad, A. & Vadgama, J. V. (2018). Combination of resveratrol and 5-flurouracil enhanced anti-telomerase activity and apoptosis by inhibiting STAT3 and Akt signaling pathways in human colorectal cancer cells. *Oncotarget 9*: 32943–32957.

de Jong, S., Timmer-Bosscha, H., de Vries, E. G. & Mulder, N. H. (1993). Effect of novobiocin on cisplatin cytotoxicity and DNA interstrand cross-link formation in a cisplatin-resistant, small-cell lung carcinoma cell line. *Int J Cancer 53*: 110–117.

Dun, J., Chen, X., Gao, H., Zhang, Y., Zhang, H. & Zhang, Y. (2015). Resveratrol synergistically augments anti-tumor effect of 5-FU in vitro and in vivo by increasing S-phase arrest and tumor apoptosis. *Exp Biol Med (Maywood) 240*: 1672–1681.

Eder, J. P., Teicher, B. A., Holden, S. A., Cathcart, K. N. & Schnipper, L. E. (1987). Novobiocin enhances alkylating agent cytotoxicity and DNA interstrand crosslinks in a murine model. *J Clin Invest 79*: 1524–1528.

Farrand, L., Byun, S., Kim, J. Y., Im-Aram, A., Lee, J., Lim, S., Lee, K. W., Suh, J. Y., Lee, H. J. & Tsang, B. K. (2013). Piceatannol enhances cisplatin sensitivity in ovarian cancer via modulation of p53, X-linked inhibitor of apoptosis protein (XIAP), and mitochondrial fission. *J Biol Chem 288*: 23740–23750.

Favaro, G., Romanello, V., Varanita, T., Andrea Desbats, M., Morbidoni, V., Tezze, C., Albiero, M., Canato, M., Gherardi, G., De Stefani, D., Mammucari, C., Blaauw, B., Boncompagni, S., Protasi, F., Reggiani, C., Scorrano, L., Salviati, L. & Sandri, M. (2019). DRP1-mediated mitochondrial shape controls calcium homeostasis and muscle mass. *Nature Communications 10*: 2576.

Hecquet, C. M., Zhang, M., Mittal, M., Vogel, S. M., Di, A., Gao, X., Bonini, M. G. & Malik, A. B. (2014). Cooperative interaction of trp melastatin channel transient receptor potential (TRPM2) with its splice variant TRPM2 short variant is essential for endothelial cell apoptosis. *Circulation research 114*: 469–479.

Hsu, C.-W., Shou, D., Huang, R., Khuc, T., Dai, S., Zheng, W., Klumpp-Thomas, C. & Xia, M. (2016). Identification of HDAC inhibitors using a cell-based HDAC I/II assay. *Journal of Biomolecular Screening 21*: 643–652.

Huang, L., Zhang, S., Zhou, J. & Li, X. (2019). Effect of resveratrol on drug resistance in colon cancer chemotherapy. *RSC Advances 9*: 2572–2580.

Huang, Y., Winkler, P. A., Sun, W., Lü, W. & Du, J. (2018). Architecture of the TRPM2 channel and its activation mechanism by ADP-ribose and calcium. *Nature 562*: 145–149.

Jiang, Q., Yang, M., Qu, Z., Zhou, J. & Zhang, Q. (2017). Resveratrol enhances anticancer effects of paclitaxel in HepG2 human liver cancer cells. *BMC Complementary and Alternative Medicine 17*: 477.

Jiang, X.-y., Zhu, X.-s., Xu, H.-y., Zhao, Z.-x., Li, S.-y., Li, S.-z., Cai, J.-h. & Cao, J.-m. (2017). Diallyl trisulfide suppresses tumor growth through the attenuation of Nrf2/Akt and activation of p38/JNK and potentiates cisplatin efficacy in gastric cancer treatment. *Acta Pharmacologica Sinica 38*: 1048–1058.

Jiang, Z., Chen, X., Chen, K., Sun, L., Gao, L., Zhou, C., Lei, M., Duan, W., Wang, Z., Ma, Q. & Ma, J. (2016). YAP inhibition by resveratrol via activation of AMPK enhances the sensitivity of pancreatic cancer cells to gemcitabine. *Nutrients 8*: 546.

Jin, X., Wei, Y., Liu, Y., Lu, X., Ding, F., Wang, J. & Yang, S. (2019). Resveratrol promotes sensitization to doxorubicin by inhibiting epithelial-mesenchymal transition and modulating SIRT1/β-catenin signaling pathway in breast cancer. *Cancer Med 8*: 1246–1257.

Kershaw, J. & Kim, K. H. (2017). The therapeutic potential of piceatannol, a natural stilbene, in metabolic diseases: a review. *Journal of Medicinal Food 20*: 427–438.

Kim, D., Lee, I. S., Jung, J. H. & Yang, S. I. (1999). Psammaplin A, a natural bromotyrosine derivative from a sponge, possesses the antibacterial activity against methicillin-resistant Staphylococcus aureus and the DNA gyrase-inhibitory activity. *Arch Pharm Res 22*: 25–29.

Kobayashi, M., Mikami, D., Uwada, J., Yazawa, T., Kamiyama, K., Kimura, H., Taniguchi, T. & Iwano, M. (2018). A short-chain fatty acid, propionate, enhances the cytotoxic effect of cisplatin by modulating GPR41 signaling pathways in HepG2 cells. *Oncotarget 9*: 31342–31354.

Kong, F., Zhang, R., Zhao, X., Zheng, G., Wang, Z. & Wang, P. (2017). Resveratrol raises in vitro anticancer effects of paclitaxel in NSCLC cell line A549 through COX-2 expression. *Korean J Physiol Pharmacol 21*: 465–474.

Kukreja, A. & Wadhwa, N. (2014). Therapeutic role of resveratrol and piceatannol in disease prevention. *Journal of Blood Disorders & Transfusion 05*: 240.

Kuo, H.-C., Lee, H.-J., hu, c. c., Shun, H.-I. & Tseng, T.-H. (2006). Enhancement of esculetin on taxol-induced apoptosis in human hepatoma HepG2 cells. *Toxicology and Applied Pharmacology 210*: 55–62.

Küpeli Akkol, E., Genç, Y., Karpuz, B., Sobarzo-Sánchez, E. & Capasso, R. (2020). Coumarins and coumarin-related compounds in pharmacotherapy of cancer. *Cancers (Basel) 12*: 1959.

Lee, S. H., Koo, B. S., Park, S. Y. & Kim, Y. M. (2015). Anti-angiogenic effects of resveratrol in combination with 5-fluorouracil on B16 murine melanoma cells. *Molecular Medicine Reports 12*: 2777–2783.

Lee, Y. H., Chen, Y. Y., Yeh, Y. L., Wang, Y. J. & Chen, R. J. (2019). Stilbene compounds inhibit tumor growth by the induction of cellular senescence and the inhibition of telomerase activity. *International Journal of Molecular Sciences 20*: 2716.

Lovitt, C. J., Shelper, T. B. & Avery, V. M. (2018). Doxorubicin resistance in breast cancer cells is mediated by extracellular matrix proteins. *BMC Cancer 18*: 41.

Mao, Q. & Unadkat, J. D. (2015). Role of the breast cancer resistance protein (BCRP/ABCG2) in drug transport – an update. *Aaps j 17*: 65–82.

Marni, R., Kundrapu, D. B., Chakraborti, A. & Malla, R. (2022). Insight into drug sensitizing effect of diallyl disulfide and diallyl trisulfide from Allium sativum L. on paclitaxel-resistant triple-negative breast cancer cells. *Journal of Ethnopharmacology 296*: 115452.

Muninathan, N. (2021). Amelioration of combination of paclitaxel and di allyl sulfide on the alterations of Bcl2, p53 and apoptosis changes against 7,12 di methyl benz (A) anthra-cene induced skin cancer in experimental animals. *Indian J Clin Biochem 36*: 143–150.

Nessa, M. U., Beale, P., Chan, C., Yu, J. Q. & Huq, F. (2012). Combinations of resveratrol, cis-platin and oxaliplatin applied to human ovarian cancer cells. *Anticancer Res 32*: 53–59.

Nordenberg, J., Kornfeld, J., Wasserman, L., Shafran, M., Halabe, E., Beery, E., Landau, O., Novogrodsky, A. & Sidi, Y. (1994). Novobiocin modulates colchicine sensitivity in par-ental and multidrug-resistant B16 melanoma cells. *Journal of Cancer Research and Clinical Oncology 120*: 599–604.

Ohira, H., Tsutsui, W. & Fujioka, Y. (2017). Are short chain fatty acids in gut microbiota defen-sive players for inflammation and atherosclerosis? *J Atheroscler Thromb 24*: 660–672.

Öztürk, Y., Günaydın, C., Yalçın, F., Nazıroğlu, M. & Braidy, N. (2019). Resveratrol enhances apoptotic and oxidant effects of paclitaxel through TRPM2 channel activation in DBTRG glioblastoma cells. *Oxid Med Cell Longev 2019*: 4619865.

Pan, Y., Lin, S., Xing, R., Zhu, M., Lin, B., Cui, J., Li, W., Gao, J., Shen, L., Zhao, Y., Guo, M., Wang, J. M., Huang, J. & Lu, Y. (2016). Epigenetic upregulation of metallothionein

2A by diallyl trisulfide enhances chemosensitivity of human gastric cancer cells to docetaxel through attenuating NF-κB activation. *Antioxid Redox Signal* 24: 839–854.

Ploner, C., Kofler, R. & Villunger, A. (2008). Noxa: at the tip of the balance between life and death. *Oncogene* 27 Suppl 1: S84–92.

Potter, G. A., Patterson, L. H., Wanogho, E., Perry, P. J., Butler, P. C., Ijaz, T., Ruparelia, K. C., Lamb, J. H., Farmer, P. B., Stanley, L. A. & Burke, M. D. (2002). The cancer preventative agent resveratrol is converted to the anticancer agent piceatannol by the cytochrome P450 enzyme CYP1B1. *Br J Cancer* 86: 774–778.

Ruhee, R. T., Roberts, L. A., Ma, S. & Suzuki, K. (2020). Organosulfur compounds: a review of their anti-inflammatory effects in human health. *Front Nutr* 7: 64.

Schwartz, G. N., Teicher, B. A., Eder, J. P., Jr., Korbut, T., Holden, S. A., Ara, G. & Herman, T. S. (1993). Modulation of antitumor alkylating agents by novobiocin, topotecan, and lonidamine. *Cancer Chemother Pharmacol* 32: 455–462.

Silva, Y. P., Bernardi, A. & Frozza, R. L. (2020). The role of short-chain fatty acids from gut microbiota in gut-brain communication. *Front Endocrinol (Lausanne)* 11: 25.

Sirerol, J. A., Rodríguez, M. L., Mena, S., Asensi, M. A., Estrela, J. M. & Ortega, A. L. (2016). Role of natural stilbenes in the prevention of cancer. *Oxid Med Cell Longev* 2016: 3128951.

Soliman, B., Farrag, A. R. & Khaled, S. (2018). Combinational effect of 5-flourouracil and resveratrol against N-Nitroso-N-Methyl urea induced colorectal cancer. *Egyptian Journal of Hospital Medicine* 70: 994–1006.

Sprouse, A. A. & Herbert, B.-S. (2014). Resveratrol augments paclitaxel treatment in MDA-MB-231 and paclitaxel-resistant MDA-MB-231 breast cancer cells. *Anticancer Res* 34: 5363–5374.

Venturelli, S., Berger, A., Böcker, A., Busch, C., Weiland, T., Noor, S., Leischner, C., Schleicher, S., Mayer, M., Weiss, T. S., Bischoff, S. C., Lauer, U. M. & Bitzer, M. (2013). Resveratrol as a pan-HDAC inhibitor alters the acetylation status of histone [corrected] proteins in human-derived hepatoblastoma cells. *PLoS One* 8: e73097–e73097.

Venugopala, K. N., Rashmi, V. & Odhav, B. (2013). Review on natural coumarin lead compounds for their pharmacological activity. *Biomed Res Int* 2013: 963248.

Xu, B. & Tao, Z. Z. (2014). Piceatannol enhances the antitumor efficacy of gemcitabine in human A549 non-small cell lung cancer cells. *Oncol Res* 22: 213–217.

Xu, J., Liu, D., Niu, H., Zhu, G., Xu, Y., Ye, D., Li, J. & Zhang, Q. (2017). Resveratrol reverses doxorubicin resistance by inhibiting epithelial-mesenchymal transition (EMT) through modulating PTEN/Akt signaling pathway in gastric cancer. *J Exp Clin Cancer Res* 36: 19.

Yang, C. H., Chen, Y. C. & Kuo, M. L. (2003). Novobiocin sensitizes BCRP/MXR/ABCP overexpressing topotecan-resistant human breast carcinoma cells to topotecan and mitoxantrone. *Anticancer Res* 23: 2519–2523.

Yang, Y., Tian, W., Yang, L., Zhang, Q., Zhu, M., Liu, Y., Li, J., Yang, L., Liu, J., Shen, Y. & Qi, Z. (2021). Gemcitabine potentiates anti-tumor effect of resveratrol on pancreatic cancer via down-regulation of VEGF-B. *Journal of Cancer Research and Clinical Oncology* 147: 903–103.

Zhou, C., Qian, W., Ma, J., Cheng, L., Jiang, Z., Yan, B., Li, J., Duan, W., Sun, L., Cao, J., Wang, F., Wu, E., Wu, Z., Ma, Q. & Li, X. (2019). Resveratrol enhances the chemotherapeutic response and reverses the stemness induced by gemcitabine in pancreatic cancer cells via targeting SREBP1. *Cell Prolif* 52: e12514.

9 Enhancing Bioactivity and Bioavailability of Natural Histone Deacetylase Inhibitors Through Innovative Nanotechnological Approach

The majority of natural histone deacetylase inhibitors (HDACi) are currently undergoing preclinical testing, few are at different stages of clinical trials and only a single natural HDAC inhibitor, romidepsin, has attained, in November 2009, FDA approval for cutaneous T-cell lymphoma, which in May 2011 was expanded to peripheral T-cell lymphoma. Three other HDACi which are synthetic have achieved this approval for subduing diverse haematological malignancies. However, the stability issue, low cancer cell accumulation, poor half-life are serious concerns still confronting the use of HDACi especially against solid tumours. Although a combinatorial approach has solved this encumbrance to a certain extent, futuristic nanotechnology-based strategies are emerging as the ray of hope for improving the stability of natural HDACi, enhancing their intra-tumoral concentration and extending their half-life in circulation. Further, the nanotechnology-backed approach improves the release kinetics of HDACi and increases their probability of on-target delivery. Moreover, nanocombinatorial approaches have proved relatively effective over the usual combined therapy in terms of preclinical as well as clinical outcome.

The preferred therapeutic efficacy of natural histone deacetylase inhibitors (HDACi) particularly against solid malignancies has still not been acquired following their use on cell lines, rodent models and human beings. The poor therapeutic efficacy of HDACi has been ascribed to the low-selectivity of HDACi, due to which they show off-target effects, inferior pharmacokinetics encompassing fast clearance and brief half-life, poor cell/tissue permeability (Scott, Haldar & Mallik 2011; Ganai 2018; Tu et al. 2020). The issues associated with HDACi-based intervention have been circumvented to a larger degree by employing trailblazing nanotechnological approaches. This tactics on the one hand makes the delivery of HDACi target specific and, on the other hand, rectifies their pharmacokinetics to an acceptable range (De Souza et al. 2020).

Various polymers and co-polymers have been employed as nanocarriers (nanovectors) for enhancing the therapeutic strength of natural HDACi. These polymers are at times linked with specific molecules like polyethylene glycol (PEG), a process known as PEGylation and with folate (Urbinati et al. 2010; Shanmugam, Rakshit & Sarkar 2022). Nanovectors have been used for loading single HDAC inhibitor or co-encapsulating HDACi with standard therapeutics. The latter approach, termed nanotechnology-mediated combinatorial therapy, often culminates in synergistic benefits and offers significant advantages over nanocarrier-based monotherapy and the usual combined therapy (Renault-Mahieux et al. 2021). These two nanotechnology-based therapeutic approaches will be discussed sequentially. As usual, first nanocarriers loaded with single HDACi will be elaborated, following which dual encapsulation will be dealt with.

9.1 NANOPARTICLES CONTAINING SINGLE NATURAL HDAC INHIBITOR

Embedding herbal HDACi in nanovectors has considerably elevated the solubility, stability and bioavailability and specificity of these promising molecules. A dramatic increase in the induction of cytotoxicity has been recorded following the use of these natural HDAC inhibitor-loaded nanoparticles. Statistically-substantial therapeutic benefits have been recorded by using these nanoformulations over the native forms of plant-isolated HDACi. These lines will become more clear by discussing some typical HDACi from plant sources in the context of nanoencapsulation.

9.1.1 NANOPARTICLES LOADED WITH QUERCETIN

This fact cannot be ignored that nano-encapsulated HDACi offer better therapeutic outcome over their free forms. Quercetin, the flavonol with propitious effect against neurodegenerative disorders and cancer, faces confrontation due to its reduced bioavailability, which coincides with its hydrophobicity. This significant concern about the bioavailability of quercetin has been subdued through nanotechnology methods. A variety of nanoparticles such as gold, magnetic nanoparticles, cyclodextrins, liposomes, polymeric nanoparticles, lipid and silica nanoparticles have been employed to address the issue of bioavailability (De Souza et al. 2020). Quercetin nanoparticles inhibited the proliferation of liver cancer cells, colony formation and migration effectually. These nanoparticles enhanced apoptosis considerably and promoted the activation of caspase-3 and 9, besides triggering the release of cytochrome c. Additionally these nanoparticles obstructed the NF-κB nuclear shuttling of NF-κB nuclear and eventually lowered cyclooxygenase 2 (COX-2) expression. Also, the suppression of Akt and ERK1/2 signalling was observed after intervention with quercetin nanoparticles (Ren et al. 2017). A copolymer composed of monomethoxy poly (ethylene glycol) and poly (ε-caprolactone) was used for encapsulating quercetin. Quercetin was loaded in these MPEG-PCL nanomicelles and, compared to free quercetin, the encapsulated form demonstrated superior effect in inhibiting the growth of colon cancer cells and committing them to apoptosis. In mice models bearing the CT26 colon tumour (subcutaneous), the polymer-encapsulated version proved to be more advanced in suppressing colon tumour growth than the non-encapsulated quercetin. Also in comparison to free quercetin, the copolymer-enclosed quercetin proved better over the usual quercetin in stimulating apoptosis, inhibiting cell proliferation and restraining angiogenesis (Xu et al. 2015). These findings certify that quercetin coated with nanosized biodegradable MPEG-PCL manifests better pharmacological effect in comparison to its free form. Hydroxyapatite nanoparticles doped with copper nanocluster (Cu-HXNPs) were used as nanocarriers for delivering quercetin to cervical cancer cells. Cu-HXNPs loaded with flavonol quercetin revealed the maximum release of quercetin under acidic environment (pH). Following its release from these nanoparticles, quercetin depleted proliferation of the aforesaid cells productively through enhancement of reactive oxygen species (ROS) production. Additionally the liberated quercetin induced apoptotic death in HeLa cells and also lowered the viability of the three-dimensional spheroids of these cells (Simon et al. 2021).

9.1.2 LUTEOLIN-LOADED NANOVECTORS IN CANCER MONOTHERAPY

This flavone modulates multiple molecular targets in different cancer types and has proved beneficial as a single agent and highly advantageous when used in conjunction with other synthetic or herbal molecules (Ganai et al. 2021). Luteolin, a flavone derived from various plant sources, was loaded in nanovesicles. Then nanovesicles embedding luteolin were generated through a solvent evaporation procedure. These nanovesicles contained cholesterol, span 60, labrasol and phosphatidylcholine in differential composition. Nonovesicles filled with luteolin were assessed for permeation, drug release, and antioxidant potential. Also, these nanovesicles were evaluated on the lung cancer line for cytotoxic effect. While these luteolin nanovesicles demonstrated negative zeta potential, their optimum polydispersibility index and size was measured to be below 0.5 and lower than 300 nm respectively. While pure luteolin showed a mean release of 20.1, the nanovesicles coated luteolin manifested maximum release (88.28) in 12 hours duration. Elevated permeation was noticed in case of encapsulated luteolin and this enhanced permeation was accredited to improved luteolin solubility due to the nanovesicles acting as surfactant. Luteolin nanovesicles evinced higher antioxidant activity and lower IC_{50} against lung cancer cells over free luteolin, signifying that luteolin containing nanovesicles offer higher therapeutic benefit over its pure form (Imam et al. 2022). Using dietary agents for vanquishing cancer has garnered maximum attention due to their wider safety window. Due to their hydrophobic character, many herbal agents demonstrate low water solubility which in reciprocation reduces their systemic delivery, bioavailability and efficacy. Nano-luteolin, which is a polymer-encapsulated form of luteolin with improved aqueous solubility, was evaluated towards the head and neck in addition to lung cancer. Nano-luteolin, in comparison to free luteolin, showed advanced growth-inhibitory effects against SCCHN, Tu21 and H292 (lung cancer line). While nano-luteolin displayed an IC_{50} of 4.13, 14.96 µM against Tu212 and H292 cells respectively, the corresponding values of luteolin were estimated to be 6.96 and 14.96 µM. Besides, experiments on mice bearing the SCCHN tumour when treated with luteolin and nano-luteolin showed that the latter has substantially potent inhibitory effect on the defined tumour over luteolin in free form (Majumdar et al. 2014). The outcome that can be derived from these results is that nano-encapsulation of luteolin by way of enhancing its solubility reduces the IC_{50} against various cancer lines and potentiates its efficacy against *in vivo* tumour-carrying models. The oral bioavailability of luteolin has also been boosted using nanoparticles. This parameter is also linked to the weak solubility of luteolin in water. Zein, a protein, and sodium caseinate which serves as a stabilizer, were employed to encapsulate flavone luteolin. Following the precipitation process the nanoparticles thus prepared were found to have negative value of zeta potential, an encapsulation efficiency of 92%, 200–300 nm as the average size and can release luteolin at pH typical to intestines. Luteolin loaded in zein nanoparticles demonstrated better free radical quenching ability than parent luteolin. Further, enhanced cytotoxic effects were recorded in case of encapsulated luteolin against colon cancer cells (Shinde et al. 2019). Conclusively, luteolin can serve as an oral drug once it is encapsulated in zein nanoparticles, which not only raises its oral bioavailability but also its cytotoxicity towards colon cancer cells. Luteolin loaded Her-2-poly (lactic-co-glycolic acid) nanoparticles were synthesized and were tested on gastric cancer cells and the underlying molecular mechanism modulated by this flavone was scrutinized. In comparison to non-targeted microspheres, markedly enhanced luteolin uptake by SGC-7901 cells was seen in the case of luteolin-loaded Her-2 nanoparticles. The loaded form of luteolin reduced the proliferation of these cells, attenuated their migration power and down-regulated the expression of FOXO1 at the transcript and at protein level (Ding et al. 2020). This all indicates that, for making the luteolin therapy targeted, the assistance of Her-2-PLGA nanoparticles is essential. Another study used pegylated bilosomes to make the therapeutic potential of luteolin better when delivered using the oral route. These luteolin-loaded PEGylated bilosomes were initially prepared through thin-film hydration and, following this, optimization was done using Box–Behnken design (software). Once the optimized formulation was sorted out, various studies including gut permeation, antioxidant and anticancer studies were performed with these optimized nanoparticles containing luteolin. The optimized luteolin-loaded pegylated bilosomes demonstrated heightened antioxidant effect in comparison to free luteolin and optimized luteolin-loaded bilosomes (non-pegylated). PEGylated bilosomes encapsulated luteolin demonstrated a propitious anticancer effect towards a couple of breast cancer lines, including MCF-7 (Zafar et al. 2021). Taking all this into consideration, it seems that PEGylated bilosomes offer superior therapeutic advantages over simple bilosomes. Another study in which luteolin was enclosed in solid lipid nanoparticles demonstrated the betterment of various properties as compared to free luteolin. The time taken to reach maximum concentration (Tmax) of luteolin loaded in solid lipid nanoparticles was quantified to be ten times lesser as compared to luteolin dispersion. The maximum concentration (Cmax) of luteolin following its administration in encapsulated form was found to be five times greater over the ordinary suspension. Luteolin in embedded form demonstrated an extended half-life, decreased clearance, distribution and improved bioavailability (Dang et al. 2014). This evidence suggests that encapsulating the luteolin flavone in solid lipid nanoparticles elevates both its aqueous solubility and plasma concentrations, which are critical for better pharmacological effect. Nanomicelles consisting of folic acid amended poly(ethylene glycol)-poly(e-caprolactone) were employed as carriers for luteolin. These micelles loaded with luteolin were named Lut/Fa-PEG-PCL and their effect was assessed on glioma even under *in vivo* conditions. Lut/Fa-PEG-PCL in comparison to the usual luteolin and luteolin encapsulated in monomethoxy poly (ethylene glycol)-poly(ε-caprolactone) designated as Lut/MPEG-PCL induced marked inhibition in the growth of glioma GL261 cells under conditions of *in vivo* as well. Further, no overt toxicity was noticed in mice subjected to either version of the luteolin (Wu et al. 2019). This signifies that modifying PEG-PCL with folate and subsequent encapsulation of luteolin in these folate-modified micelles escalates the cytotoxic effects of luteolin in glioma models without imparting a noticeable toxicity.

9.1.3 RESVERATROL NANOPARTICLES IN SINGLET CANCER THERAPY

This well-known stilbene has proved that it has a variety of effects, including an anticancer effect. However, its pharmacological benefits are restricted due to its poor solubility in water, low stability and bioavailability. PLGA nanoparticles containing resveratrol were synthesized through a solvent displacement procedure and, after eventual characterization, were evaluated on prostate cancer cells. Resveratrol-PLGA exerted a marked decline in the viability of these (LNCaP) cells as indicated by mean IC_{50} and IC_{90} of 15.6 and 41.1 μM respectively. Nanoparticles enclosing resveratrol were found to induce substantial heightened cytotoxic effect in these cells as compared to resveratrol in free mode. Importantly, the pernicious cytotoxicity was not noticed when resveratrol-loaded nanoparticles were studied using murine macrophages even at a concentration of 200 μM cells (Nassir et al. 2018). It is perceptible from these results that encapsulated resveratrol offers more advantages over free resveratrol in eliciting cytotoxic effects in prostate cancer cells while imparting no detrimental toxicity to normal cells. An interesting study where resveratrol was trapped in mesoporous nanoparticles of silica concluded that this tactic not only enhances the solubility of this stilbene but also improves its release kinetics in an *in vitro* environment. While testing these mesoporous silica nanoparticles confining resveratrol on human colon cancer lines it was evident that resveratrol in encapsulated mode showed raised cytoxicity over the free form of resveratrol. Resveratrol nanoparticle-invoked death was mediated by various pathways including PARP (Summerlin et al. 2016). Resveratrol encapsulating polymeric nanoparticles prepared through the nanoprecipitation protocol were investigated on multiple prostate cancer lines, including LNCaP, DU-145 and PC-3. Loaded nanoparticles with resveratrol showed better intracellular import and, as compared to pure resveratrol, induced better cytotoxic effect in these cells (Sanna et al. 2013). Therapeutics against hepatic carcinoma often target the entire liver and this emphasizes the requirement of therapies directed only towards the abnormal region. The rationale that avidin or biotin-attached nanoparticles show an inclination of accumulation in tumours was employed for developing chitosan nanoparticles embedding resveratrol. The other two versions of resveratrol nanoparticles with the chitosan surface modified only by biotin or by a biotin–avidin combination were also generated. While checking the pharmacokinetics of these surface-modified nanoparticles, it was deduced that avidin–biotin-modified nanoparticles quickly amassed in the liver following injection, while those linked to just biotin hampered their tendency to target liver. The anticancer activity of biotin-linked resveratrol-loaded chitosan nanoparticles and avidin–biotin-linked resveratrol-filled chitosan nanoparticles was substantially superior over resveratrol-embedding chitosan nanoparticles and solutions of trans-resveratrol. Notably, avidin–biotin-linked chitosan nanoparticles enclosing resveratrol exerted a higher cytotoxicity than just biotin-attached nanoparticles (Bu et al. 2013). This indicates that linking resveratrol containing chitosan nanoparticles to avidin–biotin drastically improves their cytotoxic effect and selectivity towards hepatic carcinoma tumour. Low stability and poor aqueous solubility as aforesaid impede the use of resveratrol for tackling aggressive cancer types. Nanoparticles composed of polyethylene glycol-polylactic acid loaded with resveratrol were examined for anticancer activity. These nutraceutical containing nanoparticles markedly reduced the number of colon cancer cells over the control. The administration of resveratrol nanoparticles intravenously in tumour-carrying mice depleted tumour growth in comparison to empty nanoparticles (Jung et al. 2015). Moreover, gelatin nanoparticles enclosing resveratrol were prepared and their anticancer effect was explored in lung cancer cells. For encapsulating resveratrol in gelatin nanoparticles glutaraldehyde, a famous cross-linking agent was used. Careful analysis demonstrated 93.6% (maximum) entrapment of resveratrol by gelatin nanoparticles. Compared to pure resveratrol, the gelatin-encapsulated form showed prompt and enhanced cellular uptake. Also, relatively, the potent inhibition of cell proliferation was noticed in the case of cells treated with resveratrol spaced in gelatin. Further, resveratrol nanoparticles induced higher ROS production, apoptosis and DNA damage when compared to its free form. Besides, stilbene enclosing nanoparticles exhibited better bioavailability and an extended half-life in comparison to resveratrol in original form (Karthikeyan et al. 2013). All this signifies that the encapsulation of stilbene resveratrol in gelatin nanoparticles is a promising strategy for its controlled delivery and towards betterment of its therapeutic index.

9.1.4 Nanoparticles Loaded with Chrysin for Cancer Treatment

Like luteolin and resveratrol, chrysin has poor aqueous solubility, due to which its biomedical advantages are impaired. Flavone chrysin also induces therapeutic effects through the inhibition of epigenetic modifying enzymes known as histone deacetylases (Ganai, Sheikh & Baba 2021). Chrysin loaded in nanoparticles composed of PLGA-PEG exhibits improved solubility, which in turn elevates its pharmacological advantages. Nanochrysin in comparison to free chrysin showed a better inhibitory effect on two breast cancer lines, including MCF-7. Importantly PLGA-PEG nanoparticles induced no cytotoxicity and were confirmed to be biocompatible, suggesting that these nanoparticles can be used in biomedical applications as nanovectors (Anari, Akbarzadeh & Zarghami 2015). Chrysin obviously has antineoplastic activity and, some years before, the impact of nanochrysin and pure chrysin on the expression of certain miRNAs was assessed in the gastric cancer cell model. The main rationale behind this study was to certify the effectiveness of nanochrysin over pure chrysin against the gastric cancer (AGS) cell line. Over free chrysin, the nanochrysin (PLGA-PEG encapsulated) showed a marked decline in

IC_{50}, which indicates the efficacy of the latter over the former. Although both chrysin and nanochrysin elevated the expression of Mir-126, Mir-22 and Mir-34a, the increase was more intense in cells treated with the encapsulated version of chrysin (Mohammadian et al. 2016). It is highly evident that nanochrysin is more powerful than ordinary chrysin towards gastric cancer cell growth inhibition. PLGA-polyvinyl alcohol nanoparticles loaded with chrysin were compared with free chrysin for a variety of therapeutic effects. Nanochrysin reduced hydrogen peroxide induced-DNA instability, and this reduction was stronger than just chrysin. When tested on two cancer cell lines, namely SKOV-3 and MCF-7, nanochrysin demonstrated better preclinical outcome than chrysin, as in the case of nanochrysin substantially alleviated IC_{50} was recorded. Nanochrysin restrained the growth of these cell lines as expected (Sulaiman, Jabir & Hameed 2018). These results certify that for deriving enhanced therapeutic advantages from flavone chrysin, it should be used in an encapsulated form in nanocarriers composed of PLGA-PVA. Nano formulations have overcome the shortcomings of natural HDACi encompassing the low availability of these molecules at target sites, non-specificity, brief half-life and, importantly, poor therapeutic index. Free chrysin has not yielded the desired results in clinical settings and this has ultimately given birth to approaches providing maximum benefit. The cytotoxic effects of chrysin-nanoparticles were unbossomed in cervical cancer cells. Nano-chrysin manifested improved inhibition than pure chrysin (Kamat, Kumari & Jayabaskaran 2022). Chrysin in its native form due to unfriendly physicochemical characteristics fails to produce the preferred results. PEG_{4000}-chrysin nanoparticles were generated through succinoyl and *cis*-aconityl linkers so that drug release specific to the tumour microenvironment can be achieved from the aforesaid PEGylated nanocarriers. Nanoparticles conjugated to chrysin through cis-aconityl and succinoyl linkers, referred to as PCNP-2 and PCNP-1 linkers respectively, were evaluated for drug release and anticancer activity. It has been observed that the profile of drug release was better in case of PCNP-2 and greater anticancer effect was also noticed in this conjugation only as compared to PCNP-1 and original chrysin. Apart from this, in comparison to native chrysin, PCNP-2 elicited a greater degree of apoptosis and more effectually curbed breast cancer cell growth (Jangid et al. 2022). The crux is that conjugating chrysin with PEG_{4000} not only rectifies its release kinetics but also improves the apoptosis induction potential to a significant level.

9.1.5 Nano-Apigenin in Cancer Therapy

Apigenin-loaded nanoparticles have been studied in hepatocellular carcinoma cells and in rats bearing this kind of carcinoma. These nanoparticles released apigenin in a sustained fashion so that it can access cancer cells in an effective manner, as well as the liver of cancer-carrying models. The progression of chemical-triggered hepatocellular carcinoma in these models was delayed on administration of apigenin nanoparticles. Following the administration of flavone-loaded nanoparticles, the concentration of apigenin escalated in the blood and liver (Bhattacharya et al. 2018). Conclusively, apigenin-filled nanoparticles significantly reduced the extremity of hepatocellular carcinoma and may prove beneficial in extending the survival of patients tolerating hepating cancer. Nanogels are emerging as important systems for drug delivery and are also used in tissue engineering. Native apigenin and apigenin-possessing nanogels were evaluated on the chronic myeloid leukaemia cell line. Gel-loaded apigenin and free apigenin-induced cytotoxicity in these cells in time- and concentration-dependent fashion. Apigenin surrounded by nanogel composed of stearate-chitosan exerted a higher cytotoxic effect over free apigenin (Samadian & Hashemi 2018). However, in comparison to the control the toxic effects induced either by native apigenin or nanogel-loaded apigenin were higher. PLGA nanoparticles loaded with apigenin were investigated against induced (radiation and chemical induced) skin tumour in mice. Apigenin-PLGA demonstrated superior effects in comparison to apigenin. This relatively heightened effect of nano-apigenin has been ascribed to the smaller size of these particles, due to which they achieve rapid mobility. These nanoparticles lowered chromosomal aberrations and tissue damage, enhanced the accumulation of ROS to induce mitochondrial apoptosis by way of modulating various molecular players (Das et al. 2013a). The central idea of these results is that PLGA nanoparticles deposited with apigenin may prove promising in abrogating skin cancer. The antiproliferative ability of apigenin-PLGA nanoparticles has been assessed using melanoma (A375) cells as models. Better results were associated with these nanoparticles due to quick mobility, smaller size and their action at specific site. Apigenin-filled PLGA nanoparticles got easily transported into cancer cells, enhanced ROS production, alleviated the activities of antioxidant enzymes, facilitated the severity of DNA damage and, finally, resulted in mitochondria-mediated apoptosis (Das et al. 2013b). The bottom line is that nanoapigenin has the potential to induce cytoxicity in melanoma cells and thus may serve as a better therapeutic to overcome melanoma. Apigenin-PLGA nanoparticles tailored with galactose were evaluated against hepatocellular carcinoma. The apoptotic effects of pure apigenin, apigenin-PLGA and apigenin-galactose-PLGA nanoparticles were scrutinized on HepG2 cells. Galactose-tailored nanoparticles manifested better entry into cells and proved to be highly cytotoxic and apoptotic than simple apigenin-PLGA and free apigenin. Further, apigenin-galactose-PLGA nanoparticles exerted higher protection against chemical-invoked hepatocellular carcinoma in rodent models. These tailored nanoparticles substantially soothed nodule formation, depleted the expression of MMP-2 and MMP-9, induced apoptosis. Further, scintigraphic imaging and histopathological examination also confirmed the superior therapeutic efficacy of galactose-tailored nanoparticles over non-tailored apigenin-PLGA nanoparticles in diethylnitrosamine-elicited hepatocellular carcinoma rat models (Ganguly et al. 2021). This means that tailoring apigenin-PLGA nanoparticles with galactose culminates in better therapeutic effects through the enhanced targetting of encapsulated apigenin to liver.

9.1.6 Anticancer Monotherapy with Rutin Nanoparticles

Among the predominant causes of cancer-related mortality all through the globe is non-small cell lung cancer. Liquid crystalline nanoparticles loaded with rutin (rutin-LCNs) were examined on the lung cancer A549 cell line. The defined nanoparticles manifested a propitious inhibitory effect on the proliferation and migration of these cells. These nanoparticles elicited apoptosis in these cells and restrained their colony formation (Paudel et al. 2021). The raised therapeutic effects of rutin-LCNs against A549 cells has been ascribed to the better aqueous solubility of encapsulated rutin over free rutin. Prenanoemulsion of this flavonol glycoside was prepared with the help of PEG and detergent tween through suitable techniques. Rutin in the said form showed nice aqueous solubility and in this prenanoemulsion form the structure of rutin remained intact. The cytotoxic activity of this prenanoemulsion was validated using the A549 cell line and it was found that when rutin is formulated with small-sized prenanoemulsion system its bioactivity gets doubled. While rutin in prenanoemulsion form exhibited an IC_{50} of 251.5 μM for colon carcinoma (Caco-2) cell line, the defined parameter for A549 line was estimated to be 154.8 μM (Hoai et al. 2020). From these experimentally determined facts it is obvious that the prenanoemulsion form of rutin exerts cytotoxicity against distinct cancer lines to a differential extent. Further, it is evident that this novel formulation stabilizes rutin and potentiates its biomedical effects. One more study encapsulated rutin in PLGA nanoparticles and assessed them in a rat model of hepatocellular carcinoma (Pandey et al. 2018). These nanoparticles were administered through the oral route in preclinical models and a significant improvement was noticed in haematological, hepatic and renal parameters. These nanoparticles markedly lowered the storm of pro-inflammatory cytokines. Nanoconjugates composed of rutin-chitosan were prepared and their cancer-antagonistic activity was checked on the breast cancer line having the speciality of being triple-negative. Substantial anticancer activity was observed following the treatment of these cells with this nanoconjugate, and this was evident from the IC_{50} of 12.5 μg/mL. The rutin-chitosan conjugate induced S-phase arrest and efficient apoptotic death in these cells (Chang et al. 2021). Thus, conjugating rutin with chitosan has the potential to become a promising strategy to circumvent bellicose breast cancer.

9.1.7 Piceatannol Nanoparticles in Singlet Cancer Therapy

Piceatannol-loaded chitosan-PLA nanoparticles were prepared through the dropping method. The effects of piceatannol nanoparticles were compared with empty chitosan-PLA ones on MCF-7, A549 and HepG2 cells. It was found that drug-loaded nanoparticles induce marginally better cytotoxicity than nanoparticles devoid of piceatannol (Dhanapal & Balaraman Ravindrran 2018). Bovine serum albumin encapsulating piceatannol nanoparticles were generated through the desolvation procedure and were further evaluated on colon cancer cells. These drug-loaded nanoparticles as compared to native piceatannol were highly productive in inhibiting the expression of HIF-1α and nuclear p65 in these cells. In the murine model, piceatannol-filled nanoparticles induced a substantial collapse of tumour dimensions compared with free piceatannol (Aljabali et al. 2020). Pleiotropic therapeutic benefits have been noticed following the use of piceatannol. Bilosome–zein formulations encapsulating this polyphenol stilbene after optimization were assessed for cytotoxic activity on lung cancer cells. Significantly lower IC_{50} was noticed with the adjusted formulation against the aforementioned cells (Alhakamy et al. 2021). Conclusively, bilosome–zein-enclosed piceatannol has advanced cytotoxic activity than usual piceatannol against A549 cells.

9.1.8 NANO-ENCAPSULATED SULFORAPHANE FOR THERAPY AGAINST CANCER

Sulforaphane, the herbal HDAC inhibitor, has proved its potential against prostate cancer and other types of cancers (Ganai 2016). Like other plant-isolated HDACi, its clinical applications are restricted due its hydrophobic character, poor gastrointestinal absorption and low availability at target location. For improving its pharmacological effects, sulforaphane was encapsulated in a co-polymer composed of monomethoxypoly (ethylene glycol) (mPEG) and poly (ε-caprolactone) (PCL). This encapsulated sulforaphane was tested on the breast cancer line of human origin. While the empty micelle induced slight cytotoxic effect on MCF-7 cells at a concentration of 1.5 mg/ml, sulforaphane-covering micelles exerted profound cytotoxic effect on this cell line (Danafar et al. 2017). Plant plasma membranes have been developed as nanocarriers. Broccoli-isolated membrane vesicles were employed to encapsulate isothiocyanate sulforaphane and these drug-encapsulated vesicles manifested an antiproliferative effect on melanoma cells. This effect was accompanied by cancer marker depletion and the elevation of aquaporin 3 (*AQP-3*). Further, in encapsulated form sulforaphane displayed better cellular uptake (Yepes-Molina & Carvajal 2021). Sulforaphane embedded in albumin nanoparticles was investigated on melanoma cells and in melanoma mice models. This study concluded that encapsulation enhances the efficacy of this isothiocyanate by improving its anticancer strength and therapeutic effect (Do et al. 2009). Gold nanoparticles loaded with sulforaphane were studied against solid tumours. After proper characterization, these drug-containing nanoparticles were tested for cytotoxic effect on a variety of cells lines including SW-620, MCF-7, Caco-2 and B16-F10. These nanoparticles were able to cause substantially greater reduction in tumour growth than the sulforaphane solution (Soni & Kohli 2019). Fe_3O_4 magnetic nanoparticles were gold coated and PEGylated were involved for the purpose of enclosing herbal HDACi sulforaphane. These PEGylated, gold-coated magnetic nanoparticles carrying sulforaphane evinced greater cytotoxicity over bare nanoparticles and individual sulforaphane against the MCF-7 (breast adenocarcinoma) cell line. In addition to this, the nanoparticle loading of sulforaphane potentiated its apoptosis-inducing ability in these cells (Izadi et al. 2015). It is quite perceptible that gold-coated magnetic nanoparticles in PEGylated form have the potential to serve as smart nanocarriers for sulforaphane.

9.2 NANOCOMBINATORIAL ANTICANCER THERAPY WITH HERBAL HDAC INHIBITORS

Simultaneous delivery of two drugs in co-encapsulated form offers additional pharmacological advantages over conventional combinatorial tactics. This is because nanoparticles can be decorated with certain molecules, such as biotin, and other molecules that make their delivery desired site specific (Lv et al. 2016). Although the loading of plant-derived HDACi has evinced superior therapeutic benefits than their corresponding ordinary forms, this benefit is further boosted when these plant molecules are encapsulated in nanoparticles along with standard anticancer therapeutics (Zhang et al. 2022). This approach, termed the nanocombinatorial approach, currently seems the climactic approach of acquiring maximum anticancer effect. Thus, in the nanocombinatorial method two or more drugs are nanoencapsulated and are thus co-delivered (Mohan, Narayanan, Sethuraman & Krishnan 2014).

9.2.1 Quercetin in Nano-Combinational Therapy

Quercetin, an herbal HDAC inhibitor, serves as a chemosensitizer and doxorubicin, a standard anticancer drug, were co-encapsulated in PEG nanoparticles, which were decorated with biotin. Additionally, the capability of quercetin to antagonize the multidrug resistance in breast cancer cells resistant to doxorubicin (MCF-7/ADR) was unravelled under *in vivo* conditions as well. Nanoparticles loaded with dual drugs demonstrated marked benefits over the usual drug combination, nanoparticles encapsulated with individual drugs or nanoparticles lacking biotin decoration in the above-mentioned cells. This happened at least partly due to the noticeable enhancement in intracellular doxorubicin concentration and its prolonged retentivity. Both the expression and activity of drug efflux protein (P-gp) was reduced following treatment with biotin-decorated nanoparticles. In the xenograft drug-resistant (MCF-7/ADR) model, the anticancer effect of doxorubicin was heightened by quercetin (Lv et al. 2016). This indicates that the biotin-receptor focussed targeting by dual-drug-encapsulated nanoparticles reinstates the sensitivity in doxorubicin-resistant abnormal cells by augmenting and sustaining cellular doxorubicin levels through direct and indirect inhibition of P-gp. PLGA nanoparticles co-encapsulating quercetin and tamoxifen were prepared and, after proper characterization, were evaluated using MCF-7 and Caco-2 cell lines. These nanoparticles showed greater cellular uptake and cytotoxic effects. Dual encapsulation manifested a nearly five-fold escalation in oral bioavailability than tamoxifen citrate and a three-fold increase than pure quercetin. This dual drug encapsulation demonstrated considerably intense tumour inhibition in comparison to individual drugs and their free combination in the breast cancer model. Administering these nanoparticles (dual encapsulated) many times through the oral route regulated angiogenesis. These nanoparticles induced no substantial oxidative stress or hepatotoxicity (Jain, Thanki & Jain 2013). This outcome suggests that quercetin–tamoxifen dual encapsulation in PLGA nanoparticles is an efficient approach towards the betterment of delivering quercetin and tamoxifen via the oral route. Polymeric microspheres embedding quercetin and paclitaxel were developed for pulmonary delivery. The rationale for this tactic was to enhance the intracellular retentivity of paclitaxel in quercetin presence through modulation of P-glycoprotein involved in drug efflux. *In vivo* studies regarding biodistribution and pharmacokinetics of these polymeric microspheres revealed their extended circulation duration and significant higher amassing in lungs (Liu et al. 2017). The essence of this study is that for optimistic delivery of hydrophobic drugs to the lungs polymeric microspheres are the novel option.

9.2.2 Luteolin–Other Molecules Dual Encapsulation

Only recently, an oil-based nanovector for co-delivery of quercetin and curcumin was developed, the main aim being the treatment of breast cancer following the intravenous route. This nanoemulsion demonstrated good encapsulation efficiency and loading of the aforesaid duo. In one type of nanoemulsion, quercetin was encapsulated; in the other, curcumin was enclosed; while in the third type of nanoemulsion, both drugs were co-loaded. All these formulations showed fast release initially and afterwards sustained release was observed for 48 hours duration. The cytotoxic effect of this dual-drug-loaded and single-drug-loaded nanoemulsions was verified on the MCF-7 cell line. The dual-loaded nanoemulsion induced synergistic cytotoxicity in these cells. Further, the cellular uptake of individual drug-loaded nanoemulsion was found to be 3.9 fold greater in contrast to their corresponding free forms (Rahman et al. 2022). This means nanoemulsions facilitate the uptake of loaded drugs by cells and this may be due to modulation of aqueous solubility once they are encapsulated. Three anticancer molecules, namely doxorubicin, quercetin and epigallocatechin gallate (EGCG), were loaded in distinct liposomal compartments so that they would encounter cancer cells one after the other (in tandem). While doxorubicin inhibits DNA replication through intercalation, quercetin and EGCG, being catalase inhibitors, heighten the oxidative stress. The cumulative effect of three different molecules results in the induction of apoptosis in cancer cells. Microbial infections associated with cancer patients are considered as secondary infections and the above-mentioned formulation was found to exhibit antimicrobial activity as well. Thus nanoemulsions loaded with doxorubicin–quercetin–EGCG serve the dual purpose (Das et al. 2019). On the one hand they induce cytotoxicity in cancer cells and, on the other hand, due to their microbicidal activity, have the potential to vanquish secondary microbial infections. A nanostructure lipid carrier (NLC), in which quercetin and irinotecan prodrug was encapsulated and following decoration with monoclonal antibody acting as agonist for TRAIL receptor-2, commonly known as conatumumab, was tested on colorectal carcinoma cells and in the xenograft mice model. The uptake of these conatumumab nanoparticles by HT-29 cells crossed the percentage of 70. Conatumumab-decorated dual-molecule-loaded NLCs exhibited substantially greater cytotoxicity over non-decorated co-loaded carriers, native drugs and nanolipid structures encapsulating drugs individually (Liu et al. 2022). A collapse of tumour growth was observed in the colorectal cancer-tolerating model, further certifying the ability of these nanoparticles to obstruct cancer.

9.2.3 Resveratrol and Standard Drug Co-encapsulation

One of the main hindrances for treating breast cancer is the resistance developed by these cells against doxorubicin. To subdue this resistance, improved PLGA nanoparticles were employed for co-encapsulation of resveratrol and doxorubicin. Resveratrol–doxorubicin containing nanoparticles delivered these drugs concomitantly in the nucleus of breast cancer cells (doxorubicin insensitive). These co-encapsulated nanoparticles exerted substantial cytotoxic effect towards doxorubicin-resistant MCF-7 and MDA-MB-231 cells. By way of inhibiting the expression of multiple proteins, including P-gp, BCRP and MRP-1, these dual-drug-enclosing nanoparticles could break the doxorubicin resistance in these lines. The stimulation of apoptosis in these cells by defined nanoparticles has been credited to BCL-2 and NF-κB down-regulation. In the mice model bearing tumour, doxorubicin and resveratrol co-loaded nanoparticles primarily delivered the encapsulated drugs to the abnormal tissue (tumour-tissue). As compared to native doxorubicin these nanoparticles substantially shrank tumour growth (doxorubicin-resistant) in the mice models carrying tumour without imparting any overt toxicity systemically (Zhao et al. 2016). The marrow of these results is that doxorubicin–resveratrol-filled nanoparticles can circumvent doxorubicin resistance in breast cancer cells that have already developed doxorubicin resistance. Head and neck cancer incidence is escalating in the Asian subcontinent due to the higher intake of alcohol and tobacco. Phytochemical resveratrol and conventional anticancer agent 5-fluorouracil were co-encapsulated in PEGylated liposome. These PEGylated liposomes serve as a delivery vehicle. The two molecules were co-loaded in the only nanoliposome. Although low concentration combinations of the two drugs were estimated to have antagonistic effect, the enhanced concentrations of resveratrol (20 or 30 μM) were found to exert synergistic effect when used in combination with the majority of the 5-fluorouracil concentrations (0.5 μM to 2 μM). Even resveratrol at 10 μM showed synergistic effect when combined with either 0.5 μM or 1 μM 5-fluorouracil (Mohan et al. 2014). This means that the proper concentrations of the two drugs which are to be co-encapsulated should be standardized for acquiring synergic cytotoxic effects on head and neck cancer cells. The major obstacle to cancer treatment is multidrug resistance. This impediment can be overcome by delivering a combination of two or more drugs in a lone nanocarrier. Resveratrol and paclitaxel were jointly encapsulated in PEGylated liposomes as a nanocombinatorial approach to deal with tumours resistant to therapeutics. Strong cytotoxicity was induced by these composite liposomes against MCF-7/Adr cells under *in vitro* conditions. Under the *in vivo* environment it was observed that bioavailability as well as retention of these drugs was improved by employing these PEGylated liposomes for co-delivery. In mice models bearing a drug-insensitive tumour, the aforesaid delivery method obstructed drug resistance without eliciting any noticeable toxicity (Meng et al. 2016). This evidence corroborates that for quashing drug resistance in tumours, the combined delivery of resveratrol and paclitaxel may prove promising. An important study used PEGylated nano-liposomes for delivering resveratrol and docetaxel jointly. This method was followed to assess the efficiency of co-encapsulation towards prostate cancer. Experiments regarding drug release certified that these drugs come out from these liposomes synchronously following sustained release kinetics. Studies of cellular uptake revealed that the above-mentioned method is effectual in intracellular delivery in comparison to other strategies. The molar ratio of doxorubicin to resveratrol in these co-loaded liposomes was kept as 1:2. Compared to a mixed formulation of two drugs, the liposome co-loaded formulation manifested markedly increased cytoxicity on cancer cells. More heightened caspase-3 activation was observed where the dual combination was given over the situation where doxorubicin was delivered alone. Balb/c nude mice tolerating PC3 tumour premierly reduced tumour growth and this was evident from parameters related to apoptosis and proliferation. Notably, meagre toxicity as well as extended survival was recorded in mice subjected to liposome co-loaded drugs than administered with the free combination of doxorubicin and resveratrol (Zhang et al. 2022). Thus, for achieving maximum therapeutic benefit from the resveratrol–doxorubicin combination, these drugs should be loaded together in nanoliposomes.

9.3 SHORTCOMINGS OF NANOPARTICLES INTEGRATED WITH HDAC INHIBITORS

Although nanoparticle-encapsulated synthetic and herbal HDACi have demonstrated significantly convincing therapeutic productivity than their corresponding native forms, still their clinical use is impaired by various challenges. It cannot be ignored that drug encapsulation in nanoparticles enhance their bioavailability and other therapeutically beneficial parameters. However, certain nanocarriers like liposomes are deftly perceived by the reticuloendothelial system and as such undergo clearance. It has been estimated that due to the aforesaid elimination, below 1% of the administered nanoparticles gain access to solid tumours (Wilhelm et al. 2016). Once the nanoparticles enter into the body of an organism, they are coated by several proteins, including those termed opsonins. This coating in turn alters the properties of these nanoparticles, causing their premature elimination. For improved efficacy, it is predominantly important that nanoparticles should remain in circulation for a prolonged period and should be able to safeguard themselves from opsonisation and subsequent phagocytosis. However, PEG coating has the potential to escape immune surveillance and thus has been utilized for preparing nanoparticles that can circulate for a longer time frame (Fam et al. 2020; Musolino et al. 2022).

The growth of tumours is facilitated by continuous neovascularization. Freshly formed vascular walls are not complete and therefore contain various fenestrations having a diameter ranging from 200 to 800 nm. Due to leaky vasculature of cancer tissues, anticancer molecules loaded in nanoparticles or micelles gain entry into these tissues, but fail to penetrate into typical tissues due to their normal vasculature (Maeda & Matsumura 2011). Additionally, the lymphatic vessels meant for drug excretion are not properly developed in tumours, due to which administered drugs are retained for longer duration. This entire phenomenon is termed enhanced permeability, and the retention effect is acronymed as the EPR effect (Maeda & Matsumura 2011; Wu 2021). For taking the benefit of this effect, the nanodrugs should have a size below 200 nm. Large-sized nanoparticles may not be allowed by the apertures of tumour vessels and thus there is no question of acting at the tumour level. Inversely, nanodrugs possessing a size below 6 nm are nimbly excreted through the kidneys and are thus unable to access the tumours (Longmire, Choyke & Kobayashi 2008; Attia et al. 2019). Nanotoxicity, the premier concern in developing drug carriers, is potentially linked to nanomaterials. Micelles and liposomes, being biocompatible, have soothed this concern, but the inorganic nanoparticles, being non-biodegradable, are not exempted from this issue. Thus, such nanoparticles, due to their inadequately reported safety, should be solely curbed for therapeutic use. It has been observed that porous silica nanoparticles demonstrate poor efficacy and due to silanol groups induce haemolysis. The aforesaid nanoparticles cause metabolic alterations, which in turn facilitate melanoma (Bharti et al. 2015; De Souza et al. 2020). Moreover, the liposomal formulations of a particular drug are more expensive than its corresponding ordinary form. For instance, the cost difference between conventional doxorubicin and its advanced liposomal formulation is $1,225. Thus, the concern of affordability is associated with nanomedicines. However, the issue of affordability has been solved to a certain degree through the introduction of generic liposomes (Bach & Elkin 2020).

Nanoformulations of HDACi, unlike their usual approved formulations, are not easily available. Most importantly the rigorous clinical studies demonstrating the safety and toxicity profile of singly nanoencapsulated HDACi or co-encapsulated HDACi are predominantly lacking, emphasizing the craving need for additional research in this unexplored area. Thus, the bottom line is that through the nanotechnology approach one can create smart HDACi, and through a nanocombinatorial strategy the activity of these HDACi can be further boosted. But copious clinical studies corroborating their safety are still required to elevate them for human use.

REFERENCES

Alhakamy, N. A., Caruso, G., Al-Rabia, M. W., Badr-Eldin, S. M., Aldawsari, H. M., Asfour, H. Z., Alshehri, S., Alzaharani, S. H., Alhamdan, M. M., Rizg, W. Y. & Allam, A. N. (2021). Piceatannol-loaded bilosome-stabilized zein protein exhibits enhanced cytostatic and apoptotic activities in lung cancer cells. *Pharmaceutics 13*: 638.

Aljabali, A. A. A., Bakshi, H. A., Hakkim, F. L., Haggag, Y. A., Al-Batanyeh, K. M., Al Zoubi, M. S., Al-Trad, B., Nasef, M. M., Satija, S., Mehta, M., Pabreja, K., Mishra, V., Khan, M., Abobaker, S., Azzouz, I. M., Dureja, H., Pabari, R. M., Dardouri, A. A. K., Kesharwani, P., Gupta, G., Dhar Shukla, S., Prasher, P., Charbe, N. B., Negi, P., Kapoor, D. N., Chellappan, D. K., Webba da Silva, M., Thompson, P., Dua, K., McCarron, P. & Tambuwala, M. M. (2020). Albumin nano-encapsulation of piceatannol enhances its anticancer potential in colon cancer via downregulation of nuclear p65 and HIF-1α. *Cancers (Basel) 12*: 113.

Anari, E., Akbarzadeh, A. & Zarghami, N. (2015). Chrysin-loaded PLGA-PEG nanoparticles designed for enhanced effect on the breast cancer cell line. *Artificial Cells, Nanomedicine, and Biotechnology 44*: 1–7.

Attia, M. F., Anton, N., Wallyn, J., Omran, Z. & Vandamme, T. F. (2019). An overview of active and passive targeting strategies to improve the nanocarriers efficiency to tumour sites. *J Pharm Pharmacol 71*: 1185–1198.

Bach, P. & Elkin, E. (2020). Cancer drug costs for a month of treatment at initial Food and Drug Administration approval. *New York, NY: Center for Health Policy & Outcomes, Memorial Sloan Kettering Cancer Center.*

Bharti, C., Nagaich, U., Pal, A. K. & Gulati, N. (2015). Mesoporous silica nanoparticles in target drug delivery system: A review. *Int J Pharm Investig 5*: 124–133.

Bhattacharya, S., Mondal, L., Mukherjee, B., Dutta, L., Ehsan, I., Debnath, M. C., Gaonkar, R. H., Pal, M. M. & Majumdar, S. (2018). Apigenin loaded nanoparticle delayed development of hepatocellular carcinoma in rats. *Nanomedicine 14*: 1905–1917.

Bu, L., Gan, L. C., Guo, X. Q., Chen, F. Z., Song, Q., Qi, Z., Gou, X. J., Hou, S. X. & Yao, Q. (2013). Trans-resveratrol loaded chitosan nanoparticles modified with biotin and avidin to target hepatic carcinoma. *Int J Pharm 452*: 355–362.

Chang, C., Zhang, L., Miao, Y., Fang, B. & Yang, Z. (2021). Anticancer and apoptotic-inducing effects of rutin-chitosan nanoconjugates in triple negative breast cancer cells. *Journal of Cluster Science 32*: 331–340.

Danafar, H., Sharafi, A., Kheiri Manjili, H. & Andalib, S. (2017). Sulforaphane delivery using mPEG-PCL co-polymer nanoparticles to breast cancer cells. *Pharm Dev Technol 22*: 642–651.

Dang, H., Meng, M. H. W., Zhao, H., Iqbal, J., Dai, R., Deng, Y. & Lv, F. (2014). Luteolin-loaded solid lipid nanoparticles synthesis, characterization, & improvement of bioavailability, pharmacokinetics in vitro and vivo studies. *Journal of Nanoparticle Research 16*: 2347.

Das, A., Konyak, P. M., Das, A., Dey, S. K. & Saha, C. (2019). Physicochemical characterization of dual action liposomal formulations: anticancer and antimicrobial. *Heliyon*
5: e02372.

Das, S., Das, J., Samadder, A., Paul, A. & Khuda-Bukhsh, A. R. (2013a). Efficacy of PLGA-
loaded apigenin nanoparticles in benzo[a]pyrene and ultraviolet-B induced skin cancer
of mice: mitochondria mediated apoptotic signalling cascades. *Food Chem Toxicol*
62: 670–680.

Das, S., Das, J., Samadder, A., Paul, A. & Khuda-Bukhsh, A. R. (2013b). Strategic formulation of apigenin-loaded PLGA nanoparticles for intracellular trafficking, DNA
targeting and improved therapeutic effects in skin melanoma in vitro. *Toxicology*
letters 223: 124–138.

De Souza, C., Ma, Z., Lindstrom, A. R. & Chatterji, B. P. (2020). Nanomaterials as potential
transporters of HDAC inhibitors. *Medicine in Drug Discovery 6*: 100040.

Dhanapal, J. & Balaraman Ravindrran, M. (2018). Chitosan/poly (lactic acid)-coated
piceatannol nanoparticles exert an in vitro apoptosis activity on liver, lung and breast
cancer cell lines. *Artif Cells Nanomed Biotechnol 46*: 274–282.

Ding, J., Li, Q., He, S., Xie, J., Liang, X., Wu, T. & Li, D. (2020). Luteolin-loading of Her-2-
poly (lactic-co-glycolic acid) nanoparticles and proliferative inhibition of gastric cancer
cells via targeted regulation of forkhead box protein O1. *J Cancer Res Ther 16*: 263–268.

Do, D., Pai, B., Rizvi, S. & Dsouza, M. (2009). Development of sulforaphane-encapsulated
microspheres for cancer epigenetic therapy. *International Journal of Pharmaceutics*
386: 114–121.

Fam, S. Y., Chee, C. F., Yong, C. Y., Ho, K. L., Mariatulqabtiah, A. R. & Tan, W. S. (2020).
Stealth coating of nanoparticles in drug-delivery systems. *Nanomaterials (Basel)*
10: 787.

Ganai, S. A. (2016). Histone deacetylase inhibitor sulforaphane: the phytochemical with
vibrant activity against prostate cancer. *Biomed Pharmacother 81*: 250–257.

Ganai, S. A. (2018). Designing isoform-selective inhibitors against classical HDACs for
effective anticancer therapy: insight and perspectives from in silico. *Curr Drug Targets*
19: 815–824.

Ganai, S. A., Sheikh, F. A. & Baba, Z. A. (2021). Plant flavone chrysin as an emerging histone
deacetylase inhibitor for prosperous epigenetic-based anticancer therapy. *Phytotherapy*
Research 35: 823–834.

Ganai, S. A., Sheikh, F. A., Baba, Z. A., Mir, M. A., Mantoo, M. A. & Yatoo, M. A. (2021).
Anticancer activity of the plant flavonoid luteolin against preclinical models of various
cancers and insights on different signalling mechanisms modulated. *Phytother Res*
35: 3509–3532.

Ganguly, S., Dewanjee, S., Sen, R., Chattopadhyay, D., Gaonkar, R. & Debnath, M. (2021).
Apigenin-loaded galactose tailored PLGA nanoparticles: a possible strategy for liver
targeting to treat hepatocellular carcinoma. *Colloids and Surfaces B: Biointerfaces*
204: 111778.

Hoai, T. T., Yen, P. T., Dao, T. T. B., Long, L. H., Anh, D. X., Minh, L. H., Anh, B. Q. &
Thuong, N. T. (2020). Evaluation of the cytotoxic effect of rutin prenanoemulsion in
lung and colon cancer cell lines. *Journal of Nanomaterials 2020*: 8867669.

Imam, S. S., Alshehri, S., Altamimi, M. A., Hussain, A., Alyahya, K. H., Mahdi,
W. A. & Qamar, W. (2022). Formulation and evaluation of luteolin-loaded
nanovesicles: in vitro physicochemical characterization and viability assessment.
ACS Omega 7: 1048–1056.

Izadi, A., Manjili, H. K., Ma'mani, L., Moslemi, E. & Mashhadikhan, M. (2015). Sulforaphane
loaded PEGylated iron oxide-gold core shell nanoparticles: a promising delivery system

for cancer therapy. *American International Journal of Contemporary Scientific Research* 2: 84–94.

Jain, A. K., Thanki, K. & Jain, S. (2013). Co-encapsulation of tamoxifen and quercetin in polymeric nanoparticles: implications on oral bioavailability, antitumor efficacy, and drug-induced toxicity. *Mol Pharm 10*: 3459–3474.

Jangid, A. K., Solanki, R., Patel, S., Medicherla, K., Pooja, D. & Kulhari, H. (2022). Improving anticancer activity of chrysin using tumor microenvironment pH-responsive and self-assembled nanoparticles. *ACS Omega 7*: 15919–15928.

Jung, K. H., Lee, J. H., Park, J. W., Quach, C. H. T., Moon, S. H., Cho, Y. S. & Lee, K. H. (2015). Resveratrol-loaded polymeric nanoparticles suppress glucose metabolism and tumor growth in vitro and in vivo. *Int J Pharm 478*: 251–257.

Kamat, S., Kumari, M. & Jayabaskaran, C. (2022). Visualizing the anti-cancer effects of chrysin nanoparticles by flow cytometry, microscopy and Fourier transform infrared spectroscopy. *Proc. SPIE 11977, Colloidal Nanoparticles for Biomedical Applications XVII*, 1197707 (3 March). https://doi.org/10.1117/12.2621529.

Karthikeyan, S., Rajendra Prasad, N., Ganamani, A. & Balamurugan, E. (2013). Anticancer activity of resveratrol-loaded gelatin nanoparticles on NCI-H460 non-small cell lung cancer cells. *Biomedicine & Preventive Nutrition 3*: 64–73.

Liu, K., Chen, W., Yang, T., Wen, B., Ding, D., Keidar, M., Tang, J. & Zhang, W. (2017). Paclitaxel and quercetin nanoparticles co-loaded in microspheres to prolong retention time for pulmonary drug delivery. *International Journal of Nanomedicine 12*: 8239–8255.

Liu, Y., Zhang, H., Cui, H., Zhang, F., Zhao, L., Liu, Y. & Meng, Q. (2022). Combined and targeted drugs delivery system for colorectal cancer treatment: conatumumab decorated, reactive oxygen species sensitive irinotecan prodrug and quercetin co-loaded nanostructured lipid carriers. *Drug Delivery 29*: 342–350.

Longmire, M., Choyke, P. L. & Kobayashi, H. (2008). Clearance properties of nano-sized particles and molecules as imaging agents: considerations and caveats. *Nanomedicine (Lond) 3*: 703–717.

Lv, L., Liu, C., Chen, C., Yu, X., Chen, G., Shi, Y., Qin, F., Ou, J., Qiu, K. & Li, G. (2016). Quercetin and doxorubicin co-encapsulated biotin receptor-targeting nanoparticles for minimizing drug resistance in breast cancer. *Oncotarget 7*: 32184–32199.

Maeda, H. & Matsumura, Y. (2011). EPR effect based drug design and clinical outlook for enhanced cancer chemotherapy. *Adv Drug Deliv Rev 63*: 129–130.

Majumdar, D., Jung, K. H., Zhang, H., Nannapaneni, S., Wang, X., Amin, A. R., Chen, Z., Chen, Z. G. & Shin, D. M. (2014). Luteolin nanoparticle in chemoprevention: in vitro and in vivo anticancer activity. *Cancer Prev Res (Phila) 7*: 65–73.

Meng, J., Guo, F., Xu, H., Liang, W., Wang, C. & Yang, X. D. (2016). Combination therapy using co-encapsulated resveratrol and paclitaxel in liposomes for drug resistance reversal in breast cancer cells in vivo. *Sci Rep 6*: 22390.

Mohammadian, F., Abhari, A., Dariushnejad, H., Nikanfar, A., Pilehvar-Soltanahmadi, Y. & Zarghami, N. (2016). Effects of chrysin-PLGA-PEG nanoparticles on proliferation and gene expression of miRNAs in gastric cancer cell line. *Iran J Cancer Prev 9*: e4190.

Mohan, A., Narayanan, S., Sethuraman, S. & Krishnan, U. M. (2014). Novel resveratrol and 5-fluorouracil coencapsulated in PEGylated nanoliposomes improve chemotherapeutic efficacy of combination against head and neck squamous cell carcinoma. *Biomed Res Int 2014*: 424239.

Musolino, E., Pagiatakis, C., Serio, S., Borgese, M., Gamberoni, F., Gornati, R., Bernardini, G. & Papait, R. (2022). The yin and yang of epigenetics in the field of nanoparticles. *Nanoscale Advances 4*: 979–994.

Nassir, A. M., Shahzad, N., Ibrahim, I. A. A., Ahmad, I., Md, S. & Ain, M. R. (2018). Resveratrol-loaded PLGA nanoparticles mediated programmed cell death in prostate cancer cells. *Saudi Pharm J 26*: 876–885.

Pandey, P., Rahman, M., Bhatt, P. C., Beg, S., Paul, B., Hafeez, A., Al-Abbasi, F. A., Nadeem, M. S., Baothman, O., Anwar, F. & Kumar, V. (2018). Implication of nano-antioxidant therapy for treatment of hepatocellular carcinoma using PLGA nanoparticles of rutin. *Nanomedicine (Lond) 13*: 849–870.

Paudel, K. R., Wadhwa, R., Tew, X. N., Lau, N. J. X., Madheswaran, T., Panneerselvam, J., Zeeshan, F., Kumar, P., Gupta, G., Anand, K., Singh, S. K., Jha, N. K., MacLoughlin, R., Hansbro, N. G., Liu, G., Shukla, S. D., Mehta, M., Hansbro, P. M., Chellappan, D. K. & Dua, K. (2021). Rutin loaded liquid crystalline nanoparticles inhibit non-small cell lung cancer proliferation and migration in vitro. *Life Sci 276*: 119436.

Rahman, M. A., Mittal, V., Wahab, S., Alsayari, A., Bin Muhsinah, A. & Almaghaslah, D. (2022). Intravenous nanocarrier for improved efficacy of quercetin and curcumin against breast cancer cells: development and comparison of single and dual drug-loaded formulations using hemolysis, cytotoxicity and cellular uptake studies. *Membranes (Basel) 12*: 713.

Ren, K.-W., Li, Y.-H., Wu, G., Ren, J.-Z., Lu, H.-B., Li, Z.-M. & Han, X.-W. (2017). Quercetin nanoparticles display antitumor activity via proliferation inhibition and apoptosis induction in liver cancer cells. *Int J Oncol 50*: 1299–1311.

Renault-Mahieux, M., Vieillard, V., Seguin, J., Espeau, P., Le, D. T., Lai-Kuen, R., Mignet, N., Paul, M. & Andrieux, K. (2021). Co-encapsulation of fisetin and cisplatin into liposomes for glioma therapy: from formulation to cell evaluation. *Pharmaceutics 13*: 790.

Samadian, N. & Hashemi, M. (2018). Effects of apigenin and apigenin-loaded nanogel on induction of apoptosis in human chronic myeloid leukemia cells. *Galen Med J 7*: e1008.

Sanna, V., Siddiqui, I. A., Sechi, M. & Mukhtar, H. (2013). Resveratrol-loaded nanoparticles based on poly(epsilon-caprolactone) and poly(D,L-lactic-co-glycolic acid)-poly(ethylene glycol) blend for prostate cancer treatment. *Mol Pharm 10*: 3871–3881.

Scott, M. D., Haldar, M. K. & Mallik, S. (2011). Natural product inhibitors and activators of histone deacetylases. *Bioactive Natural Products*: 273–309.

Shanmugam, G., Rakshit, S. & Sarkar, K. (2022). HDAC inhibitors: targets for tumor therapy, immune modulation and lung diseases. *Translational Oncology 16*: 101312.

Shinde, P., Agraval, H., Singh, A., Yadav, U. C. S. & Kumar, U. (2019). Synthesis of luteolin loaded zein nanoparticles for targeted cancer therapy improving bioavailability and efficacy. *Journal of Drug Delivery Science and Technology 52*: 369–378.

Simon, A. T., Dutta, D., Chattopadhyay, A. & Ghosh, S. S. (2021). Quercetin-loaded luminescent hydroxyapatite nanoparticles for theranostic application in monolayer and spheroid cultures of cervical cancer cell line in vitro. *ACS Applied Bio Materials 4*: 4495–4506.

Soni, K. & Kohli, K. (2019). Sulforaphane-decorated gold nanoparticle for anti-cancer activity: in vitro and in vivo studies. *Pharm Dev Technol 24*: 427–438.

Sulaiman, G. M., Jabir, M. S. & Hameed, A. H. (2018). Nanoscale modification of chrysin for improved of therapeutic efficiency and cytotoxicity. *Artif Cells Nanomed Biotechnol 46*: 708–720.

Summerlin, N., Qu, Z., Pujara, N., Sheng, Y., Jambhrunkar, S., McGuckin, M. & Popat, A. (2016). Colloidal mesoporous silica nanoparticles enhance the biological activity of resveratrol. *Colloids Surf B Biointerfaces 144*: 1–7.

Tu, B., Zhang, M., Liu, T. & Huang, Y. (2020). Nanotechnology-Based Histone Deacetylase Inhibitors for Cancer Therapy. *Front Cell Dev Biol 8*: 400.

Urbinati, G., Marsaud, V., Plassat, V., Fattal, E., Lesieur, S. & Renoir, J. M. (2010). Liposomes loaded with histone deacetylase inhibitors for breast cancer therapy. *Int J Pharm 397*: 184–193.

Wilhelm, S., Tavares, A. J., Dai, Q., Ohta, S., Audet, J., Dvorak, H. F. & Chan, W. C. W. (2016). Analysis of nanoparticle delivery to tumours. *Nature Reviews Materials 1*: 16014.

Wu, C., Xu, Q., Chen, X. & Liu, J. (2019). Delivery luteolin with folacin-modified nanoparticle for glioma therapy. *International Journal of Nanomedicine 14*: 7515–7531.

Wu, J. (2021). The enhanced permeability and retention (EPR) effect: the significance of the concept and methods to enhance its application. *J Pers Med 11*: 771.

Xu, G., Shi, H., Ren, L., Gou, H., Gong, D., Gao, X. & Huang, N. (2015). Enhancing the anti-colon cancer activity of quercetin by self-assembled micelles. *International Journal of Nanomedicine 10*: 2051–2063.

Yepes-Molina, L. & Carvajal, M. (2021). Nanoencapsulation of sulforaphane in broccoli membrane vesicles and their in vitro antiproliferative activity. *Pharm Biol 59*: 1490–1504.

Zafar, A., Alruwaili, N. K., Imam, S. S., Alsaidan, O. A., Yasir, M., Ghoneim, M. M., Alshehri, S., Anwer, M. K., Almurshedi, A. S. & Alanazi, A. S. (2021). Development and evaluation of luteolin loaded pegylated bilosome: optimization, in vitro characterization, and cytotoxicity study. *Drug Delivery 28*: 2562–2573.

Zhang, L., Lin, Z., Chen, Y., Gao, D., Wang, P., Lin, Y., Wang, Y., Wang, F., Han, Y. & Yuan, H. (2022). Co-delivery of docetaxel and resveratrol by liposomes synergistically boosts antitumor efficiency against prostate cancer. *Eur J Pharm Sci 174*: 106199.

Zhao, Y., Huan, M. L., Liu, M., Cheng, Y., Sun, Y., Cui, H., Liu, D. Z., Mei, Q. B. & Zhou, S. Y. (2016). Doxorubicin and resveratrol co-delivery nanoparticle to overcome doxorubicin resistance. *Sci Rep 6*: 35267.

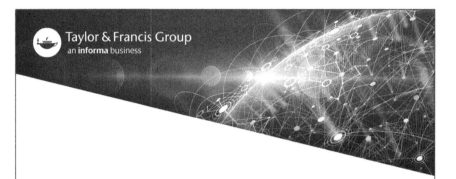

Printed in the United States
by Baker & Taylor Publisher Services